AF838529

Neurolinguistic Approach to Reading

A Guide for Speech-Language Pathologists Treating Dyslexia

Neurolinguistic Approach to Reading

A Guide for Speech-Language Pathologists Treating Dyslexia

Carol A. Kamara, PhD, CCC-SLP/A, FAAA

Plural Publishing Inc.

PLURAL PUBLISHING
INC.

5521 Ruffin Road
San Diego, CA 92123

e-mail: info@pluralpublishing.com
Website: http://www.pluralpublishing.com

FSC
www.fsc.org
MIX
Paper from responsible sources
FSC® C011935

Typeset in 11/13 Palatino by Flanagan's Publishing Services, Inc.
Printed in the United States of America by McNaughton and Gunn, Inc.

Library of Congress Cataloging-in-Publication Data

Kamara, Carol A., author.
Neurolinguistic approach to reading : a guide for speech-language pathologists treating dyslexia / Carol A. Kamara.
p. ; cm.
Includes bibliographical references and index.
ISBN 978-1-59756-655-1 (alk. paper)—ISBN 1-59756-655-1 (alk. paper)
I. Title.
[DNLM: 1. Dyslexia—physiopathology. 2. Dyslexia—rehabilitation. 3. Neurophysiology—methods. 4. Psycholinguistics—methods. 5. Reading. WL 340.6]
RC394.W6
616.85'53—dc23

2014039248

Contents

Preface

Forty years ago, I was attending a meeting for executives of speech and hearing programs in Ohio, and we were discussing the national headlines that were saying the government was going to be focusing on the literacy problem in the United States. I casually said to my colleague, "The government can pour all the money they have into the problem, but it won't get solved until speech-language pathologists become involved." My friend, ever the entrepreneur, said, "I have no idea what that means, but I like the sound of it."

Unfortunately, over the years, I have been shown to be more correct than I even knew I was at the time. Dyslexia has been the target of mountains of research, and yet thousands of little children—and big ones—have come into the offices of The Kamara Center with undiagnosed dyslexia in fourth grade, ninth grade, and even college. When parents have told schools that something must be wrong, they have been given an endless list of explanations—"The student has to be 2 years behind in reading"; "The student has to be 3 years behind in reading"; "The student isn't available for learning"; "The child has cognitive limitations"; "The child is getting good grades, so the problem can't be significant"; "The child lacks fluency, but he understands what he reads, so it's fine"; "The child lacks comprehension but reads fluently, so the child will improve"; "You should see how severe other children I see are"; "I only work on oral language"; "We don't label children"; "The student is receiving educational benefit, and that is all we are required to provide by law"; "I already have a caseload of 80"; "The student is gifted, so how could he have dyslexia?" and "I don't believe in dyslexia."

My professional experience as a clinician and president of a number of state and national associations has given me opportunity to work with many speech-language pathologists who have shared similar perspectives. Lack of speech-language pathology involvement in diagnosis and treatment of dyslexia is not a local phenomenon, and we can do better than this. We need to seriously address dyslexia and remove the barriers, many of them attitudinal. I hope that the Neurolinguistic Approach to Reading is a useful tool for all those dedicated clinicians who also know that children and adults with dyslexia need and will benefit from our help.

Introduction

What is NAR?

The Neurolinguistic Approach to Reading (NAR) is a broad-based approach to remediating dyslexia. It begins with understanding the well-documented research finding that phonologic perception skills are critical to reading acquisition and that improvement in those skills will improve reading. Even with application of this knowledge to written language acquisition, the road to literacy is not a direct path. Many factors correlate with dyslexia and must be taken into consideration. Children will have their own sets of factors that relate to their struggle to read. NAR begins with a focus on listening skills and moves to spelling, which facilitates reading, the flip side of spelling. NAR materials include a description of the four steps of the program, numerous example exercises, and forms that can be used in exercises. The program has been developed and used by the author at The Kamara Center (TKC) for more than a decade, allowing opportunity to consider outcomes longitudinally and make revisions in NAR to make it more comprehensive. NAR is geared to the speech-language pathologist (SLP) because it draws on oral and written language skills unique to that profession.

Acknowledgments

NAR has been a work in progress for over a decade. The efforts of many people have calibrated different aspects of the program. From the inception of the project to completion, the meticulous entry, merging, purging, tracking of usage changes to the spelling lists, and formatting have been carried out by Susannah Sykes. She also provided expert editing and content suggestions to the project. Her bachelor's degree in speech-language pathology and her careful, thorough approach to tasks were the source of many insights. Heather Kamara Dadmanesh used her art degree and creativity to prepare the art work and layout. She also advised project development, scouted for need resources, and edited the project as it evolved. Ramin Dadmanesh was kind enough to offer his professional editing background and his career as an English Language Learner teacher to review the document. Information technology has changed a great deal since this project began, and Allan Kamara, along with Susannah and Heather, guided our computer utilization. Dozens of clinicians over the years have used different aspects of the program, and I am grateful for their suggestions and support. To the many hundreds of children and adults who have benefitted from the program, I am most grateful for your diligence and gratified by the positive written language outcomes that NAR nudged. To my family, Allan, Heather, Michael, and David and all of their families, thank you for your patience and love as I pursued this project and many others in my career.

Dedicated to David, our dear son, brother, and Sky's Daddy

1

Provider and Recipient of NAR

Qualifications of the Speech-Language Pathologist

Who might be interested in the NAR program?

Although the Neurolinguistic Approach to Reading (NAR) program is written for speech-language pathologists (SLPs), many other professionals and parents know a great deal about language. Some psychologists, for instance, have deep insight into the relationship between language and intelligence and speak with expertise about language. Some parents of children with dyslexia are teachers or linguists and understand much about linguistic rule systems within and across languages. And many parents and teachers have been so dedicated in helping their own children with dyslexia that they have dug deeper into the research and tried different strategies, to the point that they have developed their own type of language expertise. Many people work with children with dyslexia. It takes thousands of hours to learn to read. The child will need many hours of reinforcement for emerging skills. Everyone can be put to good work, but it is critical that the plan be well coordinated across all caregivers and service providers so that the child will not be confused. Ideally, school instruction would postpone phonics instruction until phonology skills can manage the letter task, but realistically, school phonics instruction typically goes on as NAR therapy continues, and even the phonics instruction benefits from the phonology focus of treatment.

What qualifications does the SLP need to use NAR?

State licensure and/or American Speech-Language-Hearing Association (ASHA) certification already reflect SLP qualifications needed to implement and gauge the clinical process with NAR. Traditional educational preparation for a master's degree (the minimum degree requirement for certification and licensure) typically includes course work in phonology, articulation, phonetics, anatomy, physiology, hearing science, psychology, statistics, and speech science, all of which directly relate to dyslexia diagnosis and treatment strategies described in NAR. The phonology rule system expands to encompass other linguistic systems and broader considerations. Course work in syntax, semantics, child language, and human communication disorders all relate to other written language issues, some of which are addressed in NAR but primarily as they pertain to phonology coding patterns.

What national guidelines regarding reading and writing are available for SLPs?

The proficiencies needed for SLPs' clinical work in written language are also addressed by Spencer, Schuele, Guillot, and Lee (2008) and in the ASHA position statement, guidelines, and technical report, "Roles and Responsibilities of Speech-Language Pathologists with Respect to Reading and Writing in Children and Adolescents," which delineates the phonology, semantic, syntactic, morphology, and pragmatic elements of written language that SLPs address in reading and writing problems (ASHA, 2001).

What is missing in dyslexia diagnosis and treatment?

Despite this wealth of information that the SLP could bring to the individual with dyslexia, one component is typically absent. That is the application of SLP expertise to written as well as oral language. Often in internship, practicum, or even job situations, the correlation between oral and written language issues becomes apparent, and clinicians find ways to apply elements of their education to address written language. In over the past dozen years, many journal articles available to the SLP have presented written language research, nudged greater participation, suggested treatment models for the school setting, and even established the position paper on written language disorder. Still, the reality is that an insufficient number of SLPs diagnose and treat written language disorder.

Why are SLPs not more involved in dyslexia?

Insufficient involvement is not a clinician competency issue. SLPs are uniquely competent to do the work, but multiple factors interfere with participation. Often administrators of SLPs do not permit them to work with written language disorder, only oral language disorder. Insurance companies typically view dyslexia as an educational responsibility and deny coverage. Sometimes the roles and responsibilities of the reading teacher, language arts teacher, and special education teacher in written language disorder are not comfortably sorted, and an SLP can be viewed as external to the remedial process. Another obstacle is SLP availability. Thousands of SLP positions cannot be filled across the United States so that, even if administrators agreed to hire more SLPs, candidates may not be available. When clinicians already have large caseloads, they may be disinclined to take on what they view as more responsibilities. The final obstacle that NAR attempts to address is lack of familiarity with clinician roles and responsibilities. Many SLPs want to work with written language problems, but they have not been exposed to a broad enough conceptualization of the relationships between oral and written language and are not sure where to begin in implementing an effective treatment plan. Yet they are but a step away from significantly improving literacy.

What is a root cause of inadequate involvement?

It is not the purpose of NAR to denigrate the private and public national, state, and local allocation of funding for language-disordered children, but better perspective must be gained on the importance of SLP involvement in the diagnosis and treatment of dyslexia. Most of these delineated obstacles are caused by inadequate funding, but the cost of lost potential will be much greater in the future than funding SLP intervention would be now, not to mention the personal cost to millions of people living with inadequately treated dyslexia. With such a great number of children struggling to read and write, could it be that our reading programs lack sufficient understanding of the intricacies of language in dyslexia that require SLP involvement?

How do SLPs address articulation disorder?

Diagnosis and treatment of articulation disorder is one of the most common practice areas in speech-language pathology. The role of voluntary and involuntary oral motor function, articulator structure, phonologic perception, and other factors must be sorted and a treatment plan established. Treatment strategies typically involve increasing awareness of articulator structure, movement, voicing, and resonance features of involved phonemes using manipulatives, descriptions, and comparisons; phonologic perception training for target versus actual phonemes; and strategies to habituate increasing gains in phoneme production during speech, including extended pronunciation of written text. SLPs are typically adept in executing this process, even if their clinical setting tends to serve clients with other types of conditions such as voice or disfluency disorders.

What do articulation disorder and dyslexia have in common?

Clinical skills used for articulation disorders are the basic skills needed to implement NAR. Of course, transfer of these principles to letter coding of phonemes represents a next level of language, but it also is a symbolic process that is based on the same principles. The SLP uses phonetic skills to objectify and clarify confusion that the individual is having in coding phonemes. The diagnostic and treatment process focuses on phoneme knowledge, even as the process evolves to encompass other linguistic rule systems, memory, and auditory processing components of the profile. For instance, as sequences of letters representing phoneme strings (words) begin to convey different syntactic and semantic relationships such as number, case, tense, and gender morphology rules, another area of SLP expertise becomes critical in sorting confusion in the individual with dyslexia. A small number of these rules, but ones that are frequently used, that require phoneme coding are included in NAR, for instance, the three pronunciation patterns of past-tense marker /-ed/ and the phonologic rules that determine which pronunciation is used. And a few print punctuation conventions not present in auditory speech are included, for instance, notation of possession or contraction functions for final [s] in words. This description represents the basic NAR program. Of course, individuals with dyslexia have facilitating and inhibiting factors unique to them that must be addressed, but that is true of every condition.

What may surprise SLPs about dyslexia?

Although SLPs know that phonology is an extremely complex linguistic rule system and routinely address error patterns children exhibit when coding sounds of the language as phonemes in oral language, they may be surprised how much the articulation disorder profile is a template for management of the phonologic perception deficit that is characteristic of dyslexia. Nonetheless, their skills to address written language disorder are well in hand and just need to be implemented.

What typically blocks longer reading and writing in children with dyslexia?

NAR addresses only the early stages of the phoneme/grapheme coding process for reading and writing, but as improvement in the phoneme coding deficit occurs, it typically allows children to read longer passages and write longer documents. As this occurs, clinicians will often discover other language vulnerabilities. They will need to address the misperceptions in reading comprehension and errors in contextual writing that involve the same array of rule confusions found in oral language, everything from regularization of irregular past-tense verbs and nouns to pragmatic errors involving use of informal vernacular forms instead of formal expression. Written language remediation has many advantages. It is relatively permanent and therefore does not have the same draw on working memory. Written notation can actually be used to document and clarify oral language errors that escape on the air that carries them. SLPs are particularly accustomed to drawing on visual-auditory connections in treatment because of the nonpermanent nature of spoken language and routinely use multisensory interactions to confirm parameters of oral language.

Dyslexia

How many children have reading problems?

Dyslexia is a puzzling behavioral phenomenon. Why can some people understand the transfer of oral language rules to written language rules and some people cannot, or do so inadequately? Dr. G. Reid Lyon, former chief of the Child Development and Behavior Branch within the National Institute of Child Health and Human Development, was responsible for the direction and management of reading research gathered and sponsored by the National Institutes of Health (NIH). Various documents discussing those findings about children struggling to read have been published over the years and made available to the public. In his contribution to *The Keys to Literacy* (Patton & Holmes, 1998, p. 8), Lyon stated that approximately 5% of children learn to read easily without formal instruction before they enter school, and another 20% to 30% of children learn to read relatively easily when introduced to formal instruction, but for approximately 60% of children, reading is more challenging, and 20% to 30% of children find reading one of the most difficult tasks that they will have to master in life.

Is dyslexia letter and word reversal?

According to the Association for Psychological Science, approximately 70% of adults, including special education teachers, university faculty, and SLPs, think that the primary feature of dyslexia is letter or word reversal, and media often perpetuate this myth (Lilienfield, Lynn, Ruscio, & Beyerstein, 2009). The article stated that, in fact, the common denominator of dyslexia is weak auditory perception of sounds that make up the words of English. After extensive analysis of the research that delineates different diagnostic features of dyslexia, the National Institute of Neurological Disorders and Stroke (NINDS) defines dyslexia simply as unexpected difficulty learning to read despite normal intelligence and motivation. The NIH has explained in various reports that weak phonologic perception is the factor that typically impedes reading acquisition, explicitly stating that dyslexia is not seeing words or letters backward. These findings can be reviewed at http://www.ninds.nih.gov/disorders/dyslexia/dyslexia.htm. This site also provides links for related publications.

What did Orton say dyslexia was due to?

The confusion about letter reversal and dyslexia is somewhat understandable, considering that neuropathologist Samuel Orton, one of the early writers on dyslexia, defined the condition in 1925 as strephosymbolia, or "turned around symbols" (Beaton, 2004, p. 179). Even Orton, however, acknowledged that the basis of the condition was not just letter reversal but was difficulty associating visual with spoken forms of words. The explanation that dyslexia is based on letter reversal may seem logical to many people. Almost all children learning the alphabet sometimes reverse letters, but eventually they outgrow the confusion. Some children with dyslexia continue to show a few or several reversals as they get older, but so do some children without dyslexia. The positioning of the "balls" and "sticks" alone in 6 of the 26 letters in the lowercase alphabet represents an arduous learning task. To this point, no study has distinguished what pattern of letter reversal coexisting with dyslexia might be significant at what age and whether it is coincidental, a comorbid condition due to some underlying factor, a subgroup of individuals with dyslexia, or just an unexplained feature that sometimes compounds the condition. However, research confirms that letter reversal is not the cause of dyslexia.

What is the best predictor of reading success?

Recognizing that literacy problems have a significant impact on school outcomes, advanced education, and meaningful employment, Lyon presented testimony to Congress, citing the need to gather and conduct more research on reading failure to improve written language skills (Lyon, 1999). Steps to clarify the reading problems facing many American children were undertaken by the National Reading Panel (NRP), which was composed of many scholars in the field of reading, and they presented their report to Congress in 2000. The total NRP project reviewed more than 100,000 studies to arrive at their conclusions, which are summarized in a free report, *Teaching Children to Read: An Evidence-Based Assessment of the Scientific Research Literature on Reading and its Implications for Reading Instruction: Reports of the Subgroups*, which is available at http://www.nichd.nih.gov/publications/nrp/smallbook.cfm. One panel studied phonologic awareness and reviewed more than 2,000 studies on phonologic perception training to find 52 studies that met more specific NRP scientific research methodology criteria. Key findings of the NRP Phonemic Awareness panel were that children's phonologic perception skills can predict reading success and that remedial reading begins with improving phonologic perception.

What predicts reading success for deaf children?

Even for deaf children, perception of phonemes is important. Colin, Magnan, Ecalle, and Leybaert (2007) note that perception of phonemes, which are auditory units of language, predicts reading success in deaf children on the same basis that it does in hearing children. Deaf children demonstrate capacity to acquire phonologic skills, and their reading improves with explicit instruction to facilitate their learning of this implicit phonologic knowledge that is not as obvious for deaf children as it is for hearing children.

What is phonological awareness?

Phonological awareness is a term often applied to simple phonologic perception tasks such as rhyming sounds, identifying first sounds in words, or counting syllables. More complex phonologic perception tasks such as isolating or manipulating phonemes to show changes in patterns is sometimes referred to as *phonemic awareness*. Schuele and Boudreau (2008, p. 6) present a chart describing increasingly complex phonologic tasks, noting that neither phonological awareness nor phonemic awareness skills should be confused with phonics. Phonics involves print, or letters, and phonological and phonemic awareness do not involve letters. NAR expands the concept of phonological and phonemic awareness far beyond routine exercises to consider the intricacies of specific phonologic confusions for each child. The SLP tracks residual misperceptions revealed in reading and spelling errors, providing specific exercises to clarify areas of phoneme merger, substitution, omission, addition, or transpositions. The clinician uses miscoding errors to send the child back to a much earlier level of phoneme perception training.

How do children with hyperlexia read?

Children with hyperlexia appear to have an innate neurobiologic language rule-processing capability for reading that minimally considers the auditory features of phoneme-grapheme coding, bypasses spelling inconsistencies or contradictions, and applies a visually cued process to pronounce words. They often read as preschoolers. They read quickly and accurately, but they show significantly weaker comprehension of what they read. Many of them have a reading deficit, but adults generally have difficulty recognizing a highly fluid reader as a child having a problem. Children with dyslexia tend to have the opposite pattern of children with hyperlexia; they comprehend significantly better than they pronounce words (orally or silently). In treating children who show either a hyperlexic or a dyslexic pattern, if they have been taught sight reading and phonics, the remedial process is more difficult. Children with dyslexia often memorize one phoneme for each letter, and children with hyperlexia pronounce letters that meet certain print criteria for pronunciation, not as words conveying meaning. Children with either condition tend to display phonologic perception deficiency in a variety of tasks, so targeting phonologic perception in treatment applies to both conditions.

The Gifts of Dyslexia

What did Geschwind observe about the talents of individuals with dyslexia?

An interesting, counterbalancing perspective on the struggle that children with dyslexia experience was presented at the "Behavioral Measures of Dyslexia" conference in 1984 by the late Norman Geschwind, a renowned psychiatrist who significantly advanced understanding of dyslexia. He stated that, hypothetically, preventing dyslexia might eliminate the condition, but it might also deprive society of greatly talented people such as artists and engineers. His theory was that brain differences between individuals with and without dyslexia accounted for reciprocal strengths and weaknesses and that the strengths of individuals with dyslexia tended to be underrecognized. He even hypothesized that cross-hemispheric cell migration, on the basis of inherent "sidedness" selectivity, might account not only for reading problems but also for the talents of individuals with dyslexia (Gray & Kavanagh, 1985, p. 17). These are not new ideas, but they also are not well understood. Certainly, many of today's large-corporation CEOs and movie stars recall their own struggles with dyslexia, which coincide with Dr. Geschwind's observations. It is interesting that a condition that is so limiting during school years flourishes for a large roster of society's "greats" once they complete their 12-year "sentences." Many who have struggled in school go on to college and manage the courses they take by leaning toward their visual or novel strengths, the arts, and outside-the-box thinking.

The Client Who Would Benefit From NAR

What behavior signals dyslexia?

The simplest way to identify the child who would benefit from NAR is to consider the NIH's definition of dyslexia—struggle learning to read without factors that would seem to account for the difficulty. Typically, struggle is not a hidden behavior in children with dyslexia. Teachers cite lagging reading levels and difficulty sounding out words. The child complains about reading and does not want to pick out a book at the library. Parents spend long, frustrating hours trying to complete reading homework. They had viewed their child as capable and had no idea that reading would be so difficult (unless they had had other experiences with dyslexia).

What is the consequence of delayed identification?

Unfortunately, by the time struggle behaviors are recognized as an indication that something is wrong, best times for identifying risk and beginning intervention have usually already passed, and now the challenge is to contain the struggle, but still, children suffer for years. Society needs to prepare a less stressful path for individuals with dyslexia, or we will—and have—paid the consequences. Dyslexia is not a small concern that some small group of influential thinkers might want to take a look at. It requires action because it affects as much as 30% of the population.

How is the severity of dyslexia matched to NAR stages?

Severity of dyslexia is often difficult to explain because the condition is so complex and can be compounded by many factors such as the child's awareness of his difference and his reaction to the awareness. Some children absorb the embarrassment so much that they withdraw, and this can become more crippling than the dyslexia. Other children with quite low scores in assessments will bounce along in treatment and seem to be unscathed by their difference. A child might need focus on just one aspect of the dyslexia—for instance, a kindergartner confused about the alphabetic principle of coding phonemes with letters because the child hears /w/ when presented the letter name of /y/. It is better to steer away from sound-letter match to simple letter:letter name match until later on and in the meantime engage in phonologic perception games. Others will need an intensive remedial program encompassing all the stages of NAR; for instance, a child in fourth grade with ninth percentile reading skills who is stuck on sight reading may need all the stages of NAR, with considerable time spent breaking down interfering strategies that have been habituated and replacing them with facilitating ones. Another fourth grader with first percentile spelling and reading scores who can barely spell any word and abandons writing almost immediately obviously has a severe problem, but so does a child with a less severe coding problem who has lost his self-esteem because of his inability to read. Highly intelligent children often have a unique response to dyslexia; their comments show that they are aware of their problem and feel betrayed—why can't **I** read. Just last year in kindergarten, a child might have been claimed genius for his three-foot LEGO display, and now other children see the child as unable to do what school is all about—reading. Depression, anxiety, withdrawal, resistive behavior, and acting out are all examples of worrisome terms parents use to describe their child when the condition becomes manifested, and mental health professionals often play a counseling role to address these concerns.

Why do adults typically need more intensive work?

The child with all low scores in phonologic perception, auditory processing, short-term memory, retrieval from long-term memory, most oral language measures, and most written language measures (including math) and oral reading fluency (not necessarily comprehension) usually is discouraged, does have severe dyslexia, and is an obvious candidate for NAR. Adults with dyslexia will also probably need all stages of NAR because they have experienced decades of using strategies that are counterproductive, for the same reasons they are counterproductive in young children, and will require the same levels of remediation as younger children do. The program allows for whatever level of implementation the individual needs. The best way to determine if and how extensively NAR activities are needed is through a solid assessment.

What premise guides NAR?

The premise of NAR is that children can learn to read and write more easily if we begin from their point of understanding instead of our adult point of understanding. This is such a critical premise. It encompasses the principle that a developmental, neurolinguistic, systematic, and individualized remedial process can facilitate reading and writing competency in children and adults with dyslexia if it begins with their perceptions and misperceptions. Every person is truly different, and remedial programs can be efficient or inefficient, some even counterproductive if not matched to individual differences.

Early Identification

What is the advantage of early intervention?

One of the goals NAR encompasses is that language learning in preschool years be more carefully examined for weaknesses and delays. For example, children with speech production errors and weak receptive vocabulary have been found to be at greatest risk for delayed phonologic awareness, which is linked to problematic reading acquisition (Rvachew & Grawburg,

2006, p. 74). A few years of preschool and very early elementary school intervention have potential to capitalize on the short window of unique language learning potential and prepare children for both oral and written language tasks. Literature is awash with advocacy for early intervention because of documented better outcomes. Therefore, NAR activities should be activated at the earliest point possible. In fact, all children would be better prepared for written language acquisition with implementation of activities described in NAR. Most 4- and 5-year-olds can easily manage the nonletter activities to ease the transition to reading, writing, and spelling, foundation skills that NAR targets.

What language behaviors in early childhood might suggest a child at risk for dyslexia?

Several checklists are available that alert to dyslexia risk, some applicable at very early ages and others once reading instruction has begun. Catts (1997) provides a helpful checklist that cites early behaviors alerting to risk such as the child's having trouble rhyming or keeping rhythm, recalling names, explaining, pronouncing words, producing speech, sequencing sounds and syllables, remembering words, and understanding auditory messages. Because dyslexia is considered a language-based disorder and SLPs are familiar with language delay symptoms, children identified as delayed in spoken language during toddler years should be considered at risk for dyslexia. Family history of dyslexia should also be considered. Checklists used for identification purposes should be assessed by a professional working with family members so that items are interpreted correctly and appropriate balance is maintained between typical developmental variations and variations that need more attention. While it is important to identify problems as early as possible, the degree of anxiety parents express about dyslexia needs to be professionally addressed as well.

What activities would help the preschooler at risk for dyslexia?

Interventions used for very young children at risk for dyslexia would be helpful for any child. For the most part, play is simply more intentional with focus on specific vulnerabilities detected. Listening activities that fine-tune sensitivity to the sounds of the language will be particularly important. Knowing the readiness skills needed for reading and strengthening them while avoiding inappropriate activities, such as memorizing letter configurations in words, can build readiness as the child transitions to later reading readiness activities.

Assessment

How does a child tell adults the basis of his reading difficulty?

Typically and unfortunately, the child with dyslexia has not been identified as a preschooler but, once in school, illuminates for adults very clearly what the deficits are if someone will just take the time to study the child. Much of the diagnostic data that would best shape an effective treatment plan will come from the manner in which the child completes a diagnostic task, not just the score. Capturing these features takes a well-prepared clinician and probably a video-recorded diagnostic session that will illuminate strengths as well as weaknesses. A minimum team needed to assess written language should be composed of a licensed SLP, a licensed psychologist, the teacher (special education, English for Speakers of Other Languages [ESOL], resource, reading specialist, etc.), and the parent(s). Other professionals and adults may provide additional input that may be relevant to specific children who have dyslexia or other conditions that accompany dyslexia. The remedial task is large, and all are welcome to help.

What accounts for the "angel" at school and the "dickens" at home?

Parents know their child best. Often they describe behaviors that may result in reprioritizing tests to be administered in the assessment. They may clarify the severity of the condition or alert to other parties who may be asked to provide additional information. A phenomenon often reported is that the child never misbehaves in school and is a model student, but when the child gets home, parents see a different child, one who shows considerable noncompliance doing reading and writing homework. The home behavior pattern seems to be a function of unconditional love that allows the child to risk the consequences of his misbehavior, counting on his not being rejected. But consider the risks that even the child must know are possible with these noncompliant behaviors. The frustration is so high that the risk seems worth it just to release the frustration and document that the task cannot be done. In fact, sometimes a teacher may not even recognize dyslexia in a very bright child who compensates and reads at grade level, and only the parent report of many hours of homework struggle and tears alerts to the possibility of dyslexia. Gifted and talented children with dyslexia face this dilemma on a regular basis, with their struggle not even detected because their weakest scores are average. The large gap in skills does not even register as significant. As Brody and Mills (1997) state in "Gifted Children With Learning Disabilities: A Review of the Issues," "It is important, however, to note that in children with high abilities, scores on any test (including processing tests) that are 'average' may be sufficient to indicate a 'deficit'" (p. 292) and that deficit affects learning.

What role does the student have in school?

The teacher is a vital source of information about the child, and since the teacher will typically be the one to implement many of the reading and writing recommendations, it is important to consider teacher perspectives. It has been this clinician's experience that children who have strongest clinical outcomes tend to be the ones whose teacher is an active and supportive member of the team, for instance, going off-site to participate in a conference with parents and outside providers or implementing a different strategy recommended by an outside special education consultant or an SLP. The school day offers a unique window on the child. The child's job is to be a student, and the community in which that job is executed is the classroom within the school. The leader of that community is the teacher, and only the teacher has an opportunity to assess how well the child executes that job in the school community in relation to all the variables. This perspective can illuminate a child's learning profile in ways that no other window can.

How do professionals' findings complement each other?

The psychologist provides an intellectual benchmark for the child, against which oral and written language performances are compared. Without this benchmark, the SLP cannot reason dynamically to account for the score profile obtained. Because intelligence tests are typically based on language, including inner language for nonverbal measures, intelligence testing results can be skewed and often require collaborative analysis of findings to capture true ability status. In addition, the psychologist can provide insight about, for instance, emotional issues, attention, executive functioning, memory, and academic skills. The SLP also examines each of these areas, either as they relate to different linguistic rule systems or as they need to be differentiated from similar-appearing conditions. For instance, the psychologist and SLP may both see a gifted child with dyslexia with Asperger syndrome (redefined as autism spectrum disorder by the American Psychiatric Association in the *Diagnostic and Statistical Manual of Mental Disorders* [Wright, 2013]) who appears to be oppositional. Emotional factors may be identified by the psychologist. The SLP may find that the nature of the autism condition plays a role in his uncooperative behavior. Children with autism have trouble understanding pragmatic discourse rules for presenting an alternate point of view and, as a result, often present negation messages by *contradicting, denying, interrupting, refusing,* or other intents (purposes of utterances) that register with the listener as too harsh. Since many language difficulties involve a combination of both emotional and linguistic weaknesses, the psychologist and SLP work together closely.

What should a comprehensive speech and language assessment include?

The SLP examines receptive and expressive oral and written language across phonologic, semantic, syntactic, prosodic, and pragmatic rule systems; various memory skills such as short-term memory, retrieval from long-term stored memory, and prospective memory; the oral mechanism for structure and function; hearing sensitivity, including middle ear function and acoustic reflexes; and auditory processing. Some SLPs work with audiologists and may defer some of the auditory testing, but hearing testing is essential for all speech and language assessments and is part of the basic protocol. Fortunately, many standardized measures are available that allow listening and hearing assessment, including central auditory processing measures, in a quiet, non-sound-treated room and are well within the training and experience levels of SLPs.

What are some advantages of comprehensive speech and language assessments?

It takes several hours to conduct this type of oral and written language evaluation, which may seem like a significant amount of time and resource expenditure, but the comprehensive speech and language assessment ultimately serves the child well because it reduces the tendency for vulnerabilities to hide and allows a targeted, efficient treatment plan. It reduces parent struggle to find answers and reduces stress on the child. A note of caution is that not all delivery settings provide the range of speech-language pathology diagnostic and treatment services that will illuminate the profile. However, every licensed SLP should have adequate background to execute a thorough assessment and provide targeted treatment based on findings. A preliminary discussion with the clinician should clarify the clinician's professional focus. Too many times, a sparse speech and language assessment misses dyslexia, and yet no other professional has the background to assess and clarify the specific language patterns that underlie dyslexia. It can be difficult to convince, for instance, an administrative hearing officer that a child needs speech and language therapy to address his dyslexia when a previous speech and language evaluation tested only a few areas and said that findings were normal. Limited testing yields limited results that may be within the normal range but are an incomplete representation of what needed to have been considered, and dyslexia can be missed. Other times, oral language deficit is acknowledged, but no mention is made of its impact on written language.

Why does remediation need to be individualized?

Testing profiles present several score patterns and test-taking behaviors commonly seen in dyslexia, for instance, significant score scatter with peaks for skills and dips for deficits. Clusters of deficits such as disorders in auditory processing, phonologic perception, and memory or a history of articulation deficit with residual dyspraxia represent common profiles. As the treatment plan that addresses each area identified as problematic is implemented, the child's response to remediation determines the progression of steps. It is a fix-what's-broken process, not a shotgun approach. Because dyslexia has many profiles, and we do not understand it well enough to mass-market treatment for one program versus another, we must be analytic in diagnosing and treating the condition. NAR is not eclectic; that is, it does not recommend bits and pieces of different approaches for remediation that sometimes even have opposing treatment perspectives that can lead to a hit-or-miss outcome. Rather, at each step of the process, NAR returns to firm developmental principles that are consistent, no matter what strategies, methodologies, and materials are used, while remaining flexible enough to allow meeting the child's specific needs. It takes clinical competency to keep intervention on track under these requirements, but if activities reflect mindfulness of the developmental nature of language and nudge skills forward in a developmental manner, it should be just a matter of time and effort until literacy emerges.

Central Auditory Processing Disorder (CAPD)

What are symptoms of CAPD?

The coding process involved in infants' learning to say words is based on what they hear in their environment. They mold the sounds they make to match the ones they hear. Some children are not skilled in perceiving the sounds of their native language. They have difficulty matching their speech to the speech of those speaking around them, and they may show their misperception in articulation problems. This misperception may involve more than just their articulation or phoneme discrimination. Many children with phonologic perception difficulty emerge with a broader range of both oral and written language difficulties. Their listening difficulties often make parents question their child's ability to hear. Parents often seek out hearing testing because hard-of-hearing children show many of the symptoms that their child displays, but parents are usually told that hearing is not the problem. As children with normal hearing but weak listening skills get older and still do not have efficient auditory perception, life can be challenging. This difficulty processing speech, or listening, can be pervasive. The children may have behaviors and conditions such as the following:

- Sensitivity to noise, for instance, startling to a vacuum cleaner
- A history of middle ear problems, even though current hearing sensitivity is normal
- Difficulty discriminating words, for instance, hearing *bread* as *bed*, *goat* as *coat*, or *doll* as *ball*
- Trouble answering questions about a story read or told to them
- Seeming inattentive or daydreaming
- An articulation problem or a history of articulation difficulty
- Trouble following directions
- Difficulty learning to read and spell
- Delayed responses in conversation

How is most instruction presented to schoolchildren?

Some of these and other behaviors are found in children with listening problems, and often these children are diagnosed with central auditory processing disorder (CAPD). As Robert Keith noted in the *SCAN-3* manual for his diagnostic central auditory processing test battery, the American National Standard Institute (ANSI), in setting classroom acoustic standards in 2002, reported that nearly two thirds of a child's school day consists of listening and participating in spoken communication (Keith, 2009, p. 44). Children with CAPD are particularly disadvantaged in listening activities and may need a variety of remedial efforts and accommodations to improve their listening skills.

What professionals typically identify CAPD?

Other conditions, such as attention-deficit disorder, share some of the same symptoms found in CAPD, so differential diagnosis is important. More complete checklists such as Children's Auditory Performance Scale (Smoski, Brunt, & Tannahill, 1998) and Fisher's (1985) Auditory Problems Checklist can be used, along with standardized and nonstandardized measures of central auditory processing skills, to clarify the listening profile. Although SLPs, psychologists, and teachers typically identify the condition, audiologists can be of assistance in testing certain components of the listening profile. Remedial activities are usually provided by SLPs because CAPD's primary impact is on oral and written language acquisition and development. The impact will need to be addressed dynamically over time, usually with auditory training embedded in remedial activities that target various language and memory issues in SLP treatment as well as in exercises that more specifically address acoustic features of signal processing, for instance, speech in noise or tone discrimination.

What condition involves trouble processing auditory information in the presence of normal hearing?

Children with CAPD have shed light on the role of listening skills in oral and written language learning. Myklebust (1954) was one of the first to distinguish children with weak listening skills from other children who show similar auditory symptoms but ultimately have different diagnoses. He refers to difficulty processing auditory information in the presence of normal hearing sensitivity as a psychoneurological learning disability

(Myklebust, 1954). Since this early work in attempting to understand the role of auditory processing deficiencies in language learning, much research has clarified the condition of CAPD. The American Academy of Audiology (AAA, 2010) provides *Clinical Practice Guidelines: Diagnosis, Treatment, and Management of Children and Adults With Central Auditory Processing Disorder*, and ASHA (2005) summarizes this research in a technical its report, *(Central) Auditory Processing Disorders*. The documents include a review of research, diagnostic procedures, symptoms, and remedial strategies that address both oral and written language deficiencies associated with CAPD.

Is CAPD the cause of dyslexia?

As more became known about auditory processing and its relationship to learning and behavior, it became clear that dyslexia and CAPD were significantly correlated. Research investigating auditory processing suggests that phoneme perception deficit is developed very early in life and has a significant effect on written language acquisition (Schulte-Korne, Deimel, Bartling, & Remschmidt, 1998, p. 340). Research continues to explore the relationship between dyslexia and CAPD. Some have hypothesized that the auditory deficit may not cause the dyslexia but occurs in association with it (Rosen, 2003). In the meantime, as more research clarifies relationships, the possibility of CAPD should be considered in the assessment of any child with a reading problem. Whether or not a child who is struggling to read and spell is diagnosed with CAPD, it is important to understand how well the child processes auditory information, especially the phonologic patterns of the language because of phonology's close tie to reading success and failure.

Where can parents go to find how other parents cope with CAPD?

A child with CAPD can present many challenges at school, at home, and in social and recreational settings. Parents who refer their child for central auditory processing assessment may find different perspectives on the condition, advertising for various commercial products, and other experiences that make interpretation of the condition and potential treatments difficult. In *The Sound of Hope*, Heymann (2010) describes how parents perceive their children with CAPD and how they find assistance. The book presents both a lay and a professional perspective on children who have significant listening problems and may be a source of guidance for understanding the condition.

Where are CAPD treatment options discussed in depth?

Various approaches and programs are available to address auditory processing weaknesses. Many of them are reviewed in *Auditory Processing Disorders* (Geffner & Ross-Swain, 2007). Chapter 18 (Medwetsky, 2007) and Chapter 19 (Burns, 2007) present reviews of various computer programs, including Fast ForWord (FFWD) developed by Scientific Learning. Martha Burns (2007) states, "FFWD may be one of the most researched language interventions ever developed" (p. 400). She concluded that original research on FFWD published in *Science* (Tallal et al., 1996) showing significant improvement in language skills after FFWD has been upheld as an accurate finding in her review of subsequent research.

Is FFWD a "silver bullet"?

Several dozen children have been administered the FFWD program(s) at The Kamara Center (TKC) over the past decade. In pre- and posttesting for these children, virtually every child showed improvement of at least one standard deviation in one or more measures, as well as other gains. Exceptions were three adolescent boys and two young children with severe neurobiologic syndromes involving agitation who did not or could not follow the protocol. TKC can therefore endorse FFWD as an effective brain training tool, with the caveat that close monitoring be provided by an SLP, with supplemental training for "plateaued" levels, incentives, and additional speech-language therapy to address problematic areas or areas not addressed in FFWD, such as understanding the physical, acoustic, and visual phoneme features associated with articulator movement. As Scientific Learning improves and adapts its programs, some of these features are being addressed, but children will typically still need more individualized and intensive work. The need for additional work is consistent with the stated purpose of FFWD, as the authors do not consider the program a silver bullet but rather a foundational boost or supplement to remedial effort.

References

American Academy of Audiology. (2010, August). *Clinical practice guidelines: Diagnosis, treatment and management of children and adults with central auditory processing disorder*. Reston, VA: Author.

American Speech-Language-Hearing Association. (2001). Roles and responsibilities of speech-language pathologists with respect to reading and writing in children and adolescents. *ASHA Supplement, 21*, 17–27.

American Speech-Language-Hearing Association. (2005). *(Central) auditory processing disorders* [Technical Report]. Retrieved from http://www.asha.org/policy doi:10.1044/policy.TR2005-00043

Beaton, A. (2004). *Dyslexia, reading and the brain*. New York, NY: Psychology Press.

Brody, L. E., & Mills, C. J. (1997). Gifted children with learning disabilities: A review of the issues. *Journal of Learning Disabilities, 7*(3), 292.

Burns, M. (2007). Application of neuroscience to remediation of auditory processing, phonological, language and reading disorders: Fast ForWord-family of products. In D. Geffner & D. Ross-Swain (Eds.), *Auditory processing disorders: Assessment, management, and treatment* (pp. 391–407). San Diego, CA: Plural.

Catts, H. W. (1997, January). The early identification of language-based reading disabilities. *Language, Speech, and Hearing Services in Schools, 28*, 86–89.

Colin, S., Magnan, A., Ecalle, J., & Leybaert, J. (2007, March). Relation between deaf children's phonological skills in kindergarten and word recognition performance in first grade. *Journal of Child Psychology and Psychiatry, 48*(2), 44.

Fisher, L. I. (1985). *Fisher's Auditory Problems Checklist*. Tampa, FL: Educational Audiology Association.

Geffner, D., & Ross-Swain, D. (Eds.). (2007). *Auditory processing disorders: Assessment, management, and treatment*. San Diego, CA: Plural.

Gray, D. B., & Kavanagh, J. F. (Eds.). (1985). *Biobehavioral measures of dyslexia*. Parkton, MD: York Press.

Heymann, L. K. (2010). *The sound of hope: Recognizing, coping with, and treating your child's auditory processing disorder*. New York, NY: Ballantine.

Keith, R. W. (2009). *SCAN-3 for children: Tests for auditory processing disorders*. San Antonio, TX: NCS Pearson.

Lilienfeld, S. O., Lynn, S. J., Ruscio, J., & Beyerstein, B. L. (2009). Myth #17: The defining feature of dyslexia is reversing letters. From *50 great myths of popular psychology: Shattering widespread misconceptions about human behavior.* Retrieved from http://www.psychologicalscience.org/media/myths/myth_17.cfm

Lyon, G. R. (1999). *Education research: Is what we don't know hurting our children?* Testimony before the Subcommittee on Basic Research, of the House Science Committee, 106th Congress. Retrieved from http://www.nrrf.org/lyons_10-26-99.htm

Medwetsky, L. (2007). Utilization of computer software as a management tool for addressing CAPD. In D. Geffner & D. Ross-Swain (Eds.), *Auditory processing disorders: Assessment, management, and treatment* (pp. 345–390). San Diego, CA: Plural.

Myklebust, H. R. (1954). *Auditory disorders in children: A manual for differential diagnosis*. New York, NY: Grune and Stratton.

Patton, S., & Holmes, M. (Eds.). (1998). *The keys to literacy*. Washington, DC: Council for Basic Education.

Rosen, S. (2003). Auditory processing in dyslexia and specific language impairment: Is there deficit? What is its nature? Does it explain anything? *Journal of Phonetics*, *31*, 524.

Rvachew, S., & Grawburg, M. (2006). Correlates of phonological awareness in preschoolers with speech sound disorders. *Journal of Speech, Language, and Hearing Research*, *49*(1), 74–87.

Schuele, C. M., & Boudreau, D. (2008, January). Phonological awareness intervention: Beyond the basics. *Language, Speech, and Hearing Services in Schools*, *39*(1), 3–20.

Schulte-Korne, G., Deimel, W., Bartling, J., & Remschmidt, H. (1998, January 26). Auditory processing and dyslexia: Evidence for a specific speech processing deficit. *Neuroreport*, *9*(2), 337–340.

Smoski, W. J., Brunt, M. A., & Tannahill, J. C. (1998). *C. H. A. P. S.: Children's Auditory Performance Scale*. Tampa, FL: Educational Audiology Association.

Spencer, E. J., Schuele, C. M., Guillot, K. M., & Lee, M. W. (2008, October). Phonemic awareness skill of speech-language pathologists and other educators. *Language, Speech, and Hearing Services in Schools*, *39*(4), 512–520.

Tallal, P., Miller, S. L., Bedi, G., Byma, G., Wang, X., Nagarajan, S. S., . . . Merzenich, M. M. (1996). Language comprehension in language-learning impaired children improved with acoustically modified speech. *Science*, *271*(5245), 81–84.

Wright, J. (2013, May). *DSM-5* redefines autism. Retrieved from http://sfari.org/news-and-opinion/blog/2013/dsm-5-redefines-autism

2

Rationale for Approach

Neurolinguistics: The Neurobiology of Language

What is the source of data for the LAD?

For at least five decades, Chomsky's theory of language acquisition has been presented as an explanation of language learning, proposing that children learn language through a genetically endowed language acquisition device (LAD). He describes the LAD as a biologically autonomous system in the brain that represents an innate knowledge of language, a genetically prewired system with which infants are born (Chomsky, 1965, pp. 30–37). The linguistic rule categories are preset and ready for input to refine the rules of the particular language to which the child is exposed. Part of Chomsky's rationale for this explanation of language learning is that the number of rules and rule contingencies needed to acquire language so far surpasses the memory and thinking capabilities of even adults who have already acquired language that it is unrealistic to think that a small child could learn the rules and their variations. Instead, he views the language environment as the source of data presented to the child's already existing LAD, which automatically and developmentally sorts rules by repeated exposure to the data. He acknowledges that this categorization process may not be unique to language and may be part of a broader cognitive sorting process, but he considers a distilled universal grammar a unique language categorization process. His evidence is that children rapidly acquire language in a highly similar developmental sequence over time and across languages. They do not learn language according to the varied circumstances of each child's language environment.

What is a nurture theory of language acquisition?

The LAD model represents a nature theory of language learning, which has been challenged by nurture theory proponents who assign language learning to a process of conditioned generalization based on environmental influences. Bates (1999, p. 2) stated that when Chomsky introduced the nature perspective of language learning, it was contrary to the widely held blank-slate view that had been based on Skinner's behavioral conditioning research. The nurture perspective is that language is learned through positive and negative reinforcement from the environment that clarifies rules.

How do some linguists propose that environment accounts for language acquisition?

Most researchers of linguistic theory agree that language is a function of neurobiologic processes, whether they hold to theories that delineate universals found across languages as Chomsky claims or as a delineation of rules and rule departures as environment shapes language. Linguists holding the latter view note that eventually rules diverge to be unique to the features of the individual and go beyond universal sentence constructs. Others who counter Chomsky's universal grammar theory view his language acquisition device as a retrofitting of rules derived by analyzing word patterns across languages.

What mechanism does Nowak offer as an evolutionary carrier of language?

Nowak (2002) approaches grammar as a computation system and formulates a mathematical description of language acquisition based on enumeration of sentences in all languages. From this process, he calculates an evolutionary genetic prototype. Although his model is roughly defined by genes, it allows for random gene changes that occur in the developmental period. He states that other environmental factors such as the universal grammar to which the child is exposed have effects on each child's ultimate universal grammar (Nowak, 2002). Tronolone (n.d.) supports an innate language learning mechanism, saying, "The human brain contains the basic structures needed to learn any language: an inborn universal grammar." He suggests that this grammar could account for the fact that mathematics and logic are universally understood across different languages and cultures.

What systems of the brain does Henry feel are involved in language?

Henry's (1995) view is that language is a biologic process that should be studied in light of the physical properties of the brain, which is the most complex human organ. He states that language can act as a chemical change because language itself is comprised of chemical changes. He claims that language patterns can change without Mendelian evolutionary requirements, even occurring within relatively short periods of time. He rejects strict localization theories of brain language centers but hypothesizes that language probably involves both cortical and limbic systems. He views the symbolic complexities of language, for instance, the metaphor, as evidence of language's capacity to evolve outside of prescribed grammar rules through semantic association.

What evidence does Everett cite that counters Chomsky's theory?

Some notable linguists, such as Mark Baker at Rutgers University, support Chomsky's theoretical perspective of a prewired capacity for language, but others such as Daniel Everett adamantly assert that Chomsky's theory of a universal grammar is not viable. Everett supports his point of view with reference to a primitive language that he has studied, which he claims does not adhere to Chomsky's universal grammar theory. In Tom Bartlett's review of Everett's work, he noted that Everett's book takes square aim at Chomsky and that the book reads like and may actually become a movie, reflecting the level of emotional investment researchers have in their perspectives of a universal grammar (Bartlett, 2012).

How does Knezek reconcile theoretical differences?

Knezek (1997) at Duke University presents a middle-ground view of nature versus nurture perspectives of language development, inherent and learned. She states that each view allows for the other, with the difference being the degree to which one of the forces is dominant in the learning process, genetic prewiring or environmental conditioning. Her position is that the mechanism accounting for efficient language rule learning may be generic to all learning, not just language learning, but she acknowledges that a preset mechanism for sorting language rules must be operational to account for the rapidity of growth and the similarities across languages. She also documents the effects of environmental circumstances on language learning. To summarize her perspectives on language acquisition, she stated, "Language . . . has proven to be the product of nature and nurture working together."

On what common theoretical principle does NAR present language reorganization?

The Neurolinguistic Approach to Reading (NAR) is based on basic language acquisition principles on which most research shows concordance, whether theories embrace a genetic acquisition device specific to language learning or a broader cognition device based on conditioned categorization. With consideration of the language processes that are operational in normal language development, consideration must be given to children who do not display a normal neurobiologic language acquisition mechanism, and that is what NAR addresses.

What limits an infant's language learning?

NAR incorporates the premise that an atypical LAD accounts for weak phonologic perception. No child could cognitively grasp the breadth of phonology patterns that are automatically synthesized with systematic exposure, yet the typical infant learns to coalesce a myriad of sounds to distinguish one phoneme and all of its allophonic variations from other phonemes and all of their allophonic variations to then say the allophonic version of that phoneme needed for the context in which it occurs. The phonologic misperceptions of the child with dyslexia suggest that coalescence is faulty, incomplete, or inefficient.

Why must clinicians recognize a nurturist point of view?

NAR also uses the nurture premise of the environmental role in language learning. Through remedial efforts that capitalize on brain change potential, the goal is to reprogram auditory processes to clarify misperceptions. This nurturist view on language learning is gleaned from clinical experience over time and research assessing the effectiveness of remedial strategies. Recognizing the power of conditioned environmental influence is essential for any remediation endeavor. As most experienced clinicians have observed, at a certain point in the remedial process, a language rule "clicks," and the child is able to generalize the rule in a manner that implies neurobiologic reorganization of that rule. For instance, an elementary school child who lisps does not need to produce without lisping every word in the *American Heritage Dictionary* that has [s] in it to be sure that the child can say an unfamiliar [s] word correctly. At a certain point, using a limited number of stimuli, the nonlisp pattern is learned and becomes automatic. The "click" of reorganization goes beyond what the clinical intervention is explicitly demonstrating and becomes generalized.

What are some factors that affect length of treatment time?

Clinicians come to depend on this generalizing phenomenon to efficiently cover complex and wide language territory within limited clinical time. That being said, some children have such pervasive rule-merger difficulties and misconceptions that each year of life presents new challenges for more novel and more advanced rule understanding that may still require clinical support over time. Fortunately, because children with dyslexia generally have average to advanced intelligence, which may or may not be captured in intelligence testing (Shaywitz, 2003), they have the cognitive resources to advance, and they generally have other compensatory strengths that allow them to succeed—if frustration does not sidetrack them.

Language: Listening ⇒ Talking ⇒ Reading ⇒ Writing

How does early oral language play a role in written language acquisition?

To consider the difficulties that some children face when acquiring written language, one must review the elements of oral language acquisition that can falter and interfere with written language acquisition. Gaining understanding of typical and atypical growth patterns in oral language allows comparison of language rule systems for children with and without written language difficulties. These comparisons can lay the foundation for remediation efforts to expedite written language acquisition.

What is the smallest unit of spoken and heard language?

Language is a symbolic process. It includes listening, speaking, reading, and writing by using auditory and visual symbols to process and express information. The smallest symbolic unit of spoken and heard language is generally considered the phoneme (sound), which is coded in writing and reading as graphemes (letters). Finger spelling, gestures, and tactile forms that are used within a community who shares that coding knowledge can supplement or substitute those gestures for oral and written communication symbols. Infants follow a developmental sequence in learning which sound patterns evolve to represent phonemes that then become symbols that are strung together to form words. That process evolves through several stages and becomes coded with letters as written language and finally emerges as the combined auditory and visual process called spelling and reading. Children with dyslexia typically show inordinate difficulty with the smallest symbolic element, the phoneme. What are readily deciphered symbols to others are not to them. Through approximation, however, they piece enough symbolic information together to arrive at varying levels of language competency.

What linguistic rule systems are adversely affected by phonologic perception problems?

Although children with dyslexia try to cope, problems within the phonology rule system affect syntactic, semantic, and pragmatic rule systems in both oral and written language. For instance, Moats (2005–2006, p. 12) states that, to avoid the struggle, poor spellers write only words that they can spell, which limits representation of their verbal power. Apel and Apel (2011) note that they cannot appropriately divide cognitive energy between punctuation, syntax, word choice, topic development, handwriting, and other skills that allow individuals to represent their ideas in writing and understand meaning in reading. They recommend careful analysis of spelling errors because they reflect the child's operational rule system. They suggest using a spontaneous writing sample initiated with a narrative prompt to display weak rule understanding and allow targeted remediation.

Reading and Writing as an Overlain Process

What neuromuscular system does speech production use?

The basic function of the mouth, nose, and throat is to support life through eating and breathing, with speech production a secondary function. The movement patterns associated with eating and breathing functions are involuntary, or automatic, and are required to sustain life. These same structures and neuromuscular systems have been overlain, or adapted, to perform speech function. The adaptation employs a voluntary, or intentional, movement process involving a different neurologic system than the one used for vegetative functions. Infants and toddlers are attempting to master voluntary control of movements in relation to mouth and throat structures and functions based on what they hear as they acquire speech. They modulate what they feel in their mouths to achieve an acoustic signal that matches what they hear in their environment. Neurologically speaking, speech represents a higher-order function than breathing and eating because it requires voluntary movement, which draws from a more advanced neurologic system.

How many coding processes must occur before letter coding?

The speech mechanism is designed to bite, chew, swallow, breathe, cough, hiccup, and grunt, and these functions continue to be primary even though they are more primitive. Any habituated involuntary pattern, such as tongue thrusting, can invade speech learning. Written language (reading, writing, and spelling) represents yet another overlain rule system that evolved from the involuntary vegetative system, to the voluntary speech system, to now represent an invented letter rule system representing speech. Written language is based on the auditory features of speech sounds produced by the mouth. Those sounds are then coded using letters. Two translation processes must occur for written language, one that regulates the eating and breathing processes to become intentional movement patterns that produce the speech phonemes, which are then sequenced as sounds in words, and a second process that translates those heard symbols into visual symbols, that is, codes the phonemes with letters.

How is spelling and punctuation yet another coding level?

To make written language even more challenging, the conventions of print spin off yet another set of rules for spelling, idea segmentation, word order, and other forms of punctuation and capitalization representing syntactic and semantic rules at a word, phrase, and sentence level. It is amazing that children ever learn to read and write! The fact that it takes 4 to 5 years of 1 to 2 hours of daily school instruction to learn the basics of reading and writing speaks to the difficulty of learning written language rules. The task is daunting.

Are oral language rules taught to children?

In a normally functioning LAD, the first overlain process (understanding and saying the sounds of a language) can be learned automatically simply by being exposed to the language used in the child's environment. The meaning and grammar of oral language rules are also learned through listening, and little didactic instruction is necessary. A parent may correct a child struggling to distinguish between two very similar phonemes by telling the child to say *rabbit*, not *wabbit*, but the child typically learns to correct *wabbit* to say *rabbit* on his own. As the child gets older, teachers tweak grammar and meaning by explicitly teaching rules, but the great majority of the job has been done automatically before the child began school.

What does stumbled speech signal?

Many children with reading and spelling difficulties have residual speech irregularities or a history of delay that maturation did not totally remedy. The slight, or not so slight, stumbling quality of many children's speech in oral reading represents a disturbance in prosody, or the rate and rhythm of speech, a condition referred to as dyspraxia. The slight pauses, groping for mouth placements to produce specific phonemes, repetitions, or phoneme transpositions reflect their struggle to match the rate and rhythm of the voiced and unvoiced air stream to the extremely rapid oral movements that are necessary to yield fluid speech. This lack of fluidity in speech is normal in children's early speech development, but it diminishes as children learn to integrate voluntary and involuntary perceptual and neuromotor processes. As young readers attempt to match their mouth movements to the letters that represent the sound in the words they are attempting to read, they may stumble, reverting to a dyspraxic fluency pattern that they may have outgrown in speaking that reemerges in reading, as they again have been pushed beyond their speaking-coding capabilities. Dyspraxia often alerts to dyslexia because a typical profile of dyslexia is slow, imprecise word pronunciation compared with significantly stronger comprehension. The dyspraxia often persists beyond early years in children with oral and written language disorder.

What can dyspraxia lead to?

Speech-language pathologists (SLPs) advocate for toddler-age intervention when children show significant dyspraxic struggle. Diane Paul-Brown, previous director of clinical issues in speech-language pathology at the American Speech-Language-Hearing Association (ASHA), and Roseanne Clausen alert parents to the fact that speech problems, including dyspraxia, "can lead to other difficulties with written language and academic and social skills," noting that the earlier the intervention, "the better the brain can organize" (Paul-Brown & Clausen, 2011). Like CAPD, a separate but related set of objectives for dyspraxia may need to be added to the treatment plan for some children. *Developmental Apraxia of Speech: Theory and Clinical Practice* provides more information on the condition by reviewing the history of apraxia research and discussing different treatment approaches (Hall, Jordan, & Robin, 1993). Ludlow et al. (2008) also provide more information in a National Institutes of Health (NIH) report on speech motor control problems, including dyspraxia, and the use of neuroimaging procedures to assess several aspects of the condition, including treatment outcomes.

The Listening Environment

At what age do babies select only the phonemes spoken in their environment?

The listening environment that allows children to differentiate between phonemes begins to have an impact on language learning during infancy or even prenatally. During early infancy, babies have the capacity to distinguish the phonemes of any language, but by 9 to 12 months of age, they become attuned to only the phonemes in their language environment. Werker and Curtin (2005, p. 200) note that infants have the ability to produce native and nonnative phonemes across languages during the babbling stage, but as they move out of the silent period following babbling, the range of phonemes in their first words contains only native phonemes. This period of reorganization coincides with the cognitive stage of categorization and may be part of a broad maturational process.

Do infants learn phonemes in a random or developmental order?

As infants begin to focus on the language they hear around them, the phonemes of speech must sound similar to what an adult perceives when listening to a foreign language. What they hear is a stream of sounds embedded on a melody and stress pattern with periodic interruptions. The stream seems to provide no clue to signal the point at which one word ends and another begins. Yet infants do find the clues, and they begin to learn phoneme patterns in a similar developmental order. Blache (1978, p. 113) notes that this phenomenon occurs across languages as he considers the basis on which phonemes are distinguished. He reviews theories that define the distinctive features of sounds that the brain uses to discriminate phonemes and notes that the child learns these phoneme-differentiating rules automatically through listening and talking with no need for explicit instruction.

Brain Plasticity

What allows young children to learn multiple languages without instruction?

Curtiss and Kuhl (1996, p. 17) state that the rate of language acquisition during the first years of life is unparalleled during any other period. The brain of a 2-year-old child has twice as many synapses as an adult's brain and is much more receptive for language learning in early years. Curtiss says that children can be exposed to and learn multiple languages at the same time during the window of early language learning years without detriment to any of the languages they are learning. With systematic exposure, they can learn each language because their brains are "ripe" to learn language. Kuhl refers to babies as "citizens of the world," ready to perceive any language they hear up to 6 months of age, when they begin to specialize in the language of their environment. This neural flexibility is more readily available from birth to approximately 6 years of age, allowing second-language learning to occur more easily. If not, the rule-sorting function becomes less plastic as the child reaches adolescence. Snedeker, Geren, and Shafto (n.d., p. 11) point out that the second-language acquisition period varies from child to child, depending on many factors such as age and exposure to the second language. In their research addressing internationally adopted children, they report that the preschoolers followed patterns of infant language development and that after 2 months, most were speaking only in English and by 1 year were reported by parents to know fewer than five words of their birth language. This neural adaptability is referred to as brain plasticity or the ability of the brain to organize and reorganize its circuitry (Hoiland, n.d.)

What light do hemispherectomy and adult research shed on brain plasticity?

All is not lost if an older child or an adult with dyslexia has passed the early neural flexibility period because brain plasticity is not exclusively an early childhood phenomenon. Even after serious injury, brain plasticity allows language relearning. Research analyzing brain plasticity in children whose medical conditions required removing half of the brain found that, even in these extreme cases, language can be relearned (Curtiss & Schaeffer, 2005, p. 163). Other researchers report significant brain changes in adults with dyslexia using neuroimaging techniques after participating in intensive reading training programs. And a series of studies addressing neuroplasticity in Japanese adults reports that subjects exposed to intensive discrimination training to distinguish phonemes [l] and [r], which their native language does not model, were significantly more successful discriminating the two phonemes (Iverson, Hazan, & Bannister, 2005; McClelland, Fiez, & McCandliss, 2002).

What four periods yield greatest brain change?

Documentation of brain plasticity comes from many arenas, and this capacity of the brain to alter synaptic structure serves as the basis for developmental as well as rehabilitative growth. The Vanderbilt Kennedy Center (n.d.) cites four periods of rapid brain change, or plasticity: (1) early development, (2) changed body function that sends different sensory information to the brain, (3) intentional change in sensory information to the brain achieved through learning and memory, and (4) brain injury. Capitalizing on the first and third periods of plasticity is important in any developmental or remedial program and are the time periods to which NAR activities are particularly applicable.

Zone of Proximal Development (ZPD)

What is the ZPD?

NAR draws on Vygotsky's zone of proximal development (ZPD) tenet that certain tasks are too difficult for children to master by themselves, but, if helped, they can master the task and achieve independent performance at the upper limit of the proximal development zone. This process allows children to draw from and maximize information in their environment to stabilize language rules. In his later work, Vygotsky (1986, p. 187) slightly modifies his tenet of step-by-step learning by stating that the zone must be considered within the limits of the child's developmental state, but the same principles of learning facilitation apply.

What does Vygotsky mean by "real communication"?

Vygotsky (1962, p. 6) notes that the phoneme is the vehicle to word meaning and the basis of real communication. For children with auditory processing difficulties that result in weak phonologic perception, as is the consistent finding for children with dyslexia, real communication in both oral and written form represents struggle. Listening and reading comprehension based on a phoneme coding system yields real communication only when a phoneme string holds meaning. If the coding system falters, comprehension suffers to varying degrees, depending on the compensatory resources available to the child. Even the child with strong compensatory strategies will register the extra effort and feel the strain it requires to engage those strategies, so the coding deficiency must be managed to minimize struggle. ZPD is the basis on which most NAR activities should be presented because it reduces stress. Tasks that are too easy produce boredom, and tasks that are too difficult produce frustration. Using ZPD avoids both problems.

How is ZPD applied to motor tasks?

An updated review of ZPD in *Simply Psychology* (McLeod, 2010/2012) notes that ZPD has become synonymous with scaffolding and describes using applications of ZPD to tennis and classroom learning situations, using principles that are universally applicable. McLeod discussed reciprocal teaching as a contemporary application of ZPD in which the adult and child collaborate by summarizing, questioning, clarifying, and predicting task elements and diminishing the adult role until the child can carry out the process independently. These are routine treatment strategies.

What happens with too much focus on steps and not the child?

The approaches needed to help children are built on inquisitiveness about where the child is, what is blocking progress, and what can reduce the obstacles. Although traditional approaches may be effective, novel approaches that emerge in a moment of discovery can also be powerful, even if they apply to only one child. Adhering to a methodology too strictly and not seeing that the child is not tracking with the methodology wastes the child's availability. NAR is based on certain step-by-step procedures but encourages adding, subtracting, or revising steps in the program, depending on the child's response. Alternative resources to expand on certain steps are often suggested. All activities should be assessed in an ongoing manner with error notes, charts, and tallies to be certain that they are meeting targets.

Brain Training

How is reprogramming accomplished?

Brain training programs are not perfect, sometimes even employing incorrect rules for specific language features such as computer programs that mispronounce phonemes or use incorrect syntax rules, but as long as adherence to developmental principles is maintained and too many glitches do not occur, the training can result in powerful changes. Children with language-learning difficulties may not have neurologic systems that are wired in a typical fashion, but that wiring can be altered with brain training. The premise is that brain plasticity allows for reprogramming of misperceived rules with systematic training so that the deficient neurologic system is rewired, rerouted, or somehow neurobiologically altered to relay accurate information necessary for language processing and production, ultimately as a neurobiochemical function.

How can individualization and brain training be combined?

Individualization of intervention is paramount in remedial programs for dyslexia. Merger of brain training, which requires intensive repetition, with individualization may seem contradictory, but both repetition and individualization must be employed in a balanced manner to nudge and stabilize gains. NAR offers examples of exercises that should be repeated many times to brain-train, but almost every child will need adaptations to the exercises, skip them, or have others added, depending on specific responses. Implementation of NAR is therefore more difficult than strict preprogramed exercises that are not alterable. Many preset programs are based on solid processing, memory, and language principles, but they may lack the adaptability that is necessary, not only to meet individual skill variances but also to take into consideration motivation variables.

How can intrinsic and extrinsic motivation be captured?

Children may call brain training *difficult*, *fun*, *too easy*, *too hard*, *boring*, *interesting*, *intense*, or *frustrating*. Responses vary, which is another reason that therapy must be individualized. Therapy targets intrinsic reward as a motivational force, but immediate gratification is also important, especially for younger children who are not yet introspective enough to assess benefit. Therefore, changes in activities, tasks with game formats, immediate recognition of correct responses to more challenging items, breaks, and even tangible rewards are needed in carrying out difficult exercises. The power of charting is often overlooked, but children respond personally to charts that register and track their performance. In fact, endurance can be built with a more adaptable approach. Playing the board game *Candyland* at intervals or after an intensive brain-training session involving phoneme sequencing erases the difficulty the child experienced—and allows analysis of game-playing and turn-taking social pragmatic skills. When all is considered, an adaptable approach achieves much more than a preset, endurance-based program, but preset programs can be useful adjuncts if used judiciously.

References

Apel, K., & Apel, L. (2011). Identifying individual differences in students' written language abilities. *Topics in Language Disorders*, *31*(1), 54–72.

Bartlett, T. (2012, March 20). *Angry words: Will one researcher's discovery deep in the Amazon destroy the foundation of modern linguistics?* Retrieved from http://chronicle.com/article/Researchers-Findings-in-the/131260/

Bates, E. (1999). *On the nature and nurture of language.* Retrieved from http://crl.ucsd.edu/bates/papers/pdf/bates-inpress.pdf

Blache, S. E. (1978). *The acquisition of distinctive features.* Baltimore, MD: University Park Press.

Chomsky, N. (1965). *Aspect of the theory of syntax.* Cambridge, MA: MIT Press.

Curtiss, S., & Kuhl, P. (1996). Research notes: Language learning and the developing brain. *Learning Languages, 1*(2), 17. Retrieved from http://www.cal.org/earlylang/benefits/research_notes.html
Curtiss, S., & Schaeffer, J. (2005). Syntactic development in children with hemispherectomy: The I-, D-, and C-systems. *Brain and Language, 94*, 147–166.
Hall, P. K., Jordan, L. S., & Robin, D. A. (1993). *Developmental apraxia of speech: Theory and clinical practice*. Austin, TX: Pro-Ed.
Henry, C. (1995). Universal grammar. *Communication and Cognition—Artificial Intelligence, 12*, (1–2), 45–61. Retrieved from http://informatics.indiana.edu/rocha/univgram.html
Hoiland, E. (n.d.). *Brain plasticity: What is it? Learning and memory. Neuroscience for kids.* Retrieved from http://www.faculty.washington.edu/chudler/plast.html
Iverson, P., Hazan, V., & Bannister, K. (2005, November). Phonetic training with acoustic cue manipulations: A comparison of methods for teaching English /r/-/l/ to Japanese adults [Abstract]. *Journal of the Acoustical Society of America, 118*(5), 3267–3278.
Knezek, M. (1997). *Nature vs. nurture: The miracle of language.* Retrieved from http://www.duke.edu/~pk10/language/psych.htm
Ludlow, C. L., Hait, J., Kent, R., Ramig, L. O., Shrivastav, R., Strand, E., . . . Sapienza, C. M. (2008, February). Translating principles of neural plasticity into research on speech motor control recovery and rehabilitation. *Journal of Speech, Language, and Hearing Research, 51*, S240–S258.
McClelland, J. L., Fiez, J. A., & McCandliss, B. D. (2002, December). Teaching the /r/-/l/ discrimination to Japanese adults: Behavioral and neural aspects. *Physiology & Behavior, 77*, 657–662.
McLeod, S. (2012). *Zone of proximal development.* Retrieved from http://www.simplypsychology.org/Zone-of-Proximal-Development.html (Original work published 2010)
Moats, L. C. (2005–2006, Winter). How spelling supports reading: And why it is more regular and predicable than you think. *American Educator*, pp. 12–43. Retrieved from http://www.aft.org/pdfs/americaneducator/winter0506/Moats.pdf
Nowak, M. A. (2002). From quasispecies to universal grammar. *Zeitschrift fur Physikalische Chemie, 216*, 5–20. Retrieved from http://www.ped.fas.harvard.edu/people/faculty/publications_nowak/ZPhysChem02.pdf
Paul-Brown, D., & Clausen, R. (2011, November). *Let's talk . . . for people with special communication needs: What is developmental apraxia of speech?* Retrieved from http://www.pediatricservices.com/parents/pc-36.htm
Shaywitz, S. (2003). *Overcoming dyslexia: A new and complete science-based program for reading problems at any level*. New York, NY: Knopf.
Snedeker, J., Geren, J., & Shafto, C. L. (n.d.). *Different paths: Changes in second-language acquisition between three and five years of age*. Retrieved from http://www.wjh.harvard.edu/~lds/pdfs/Snedeker%20Geren%20Shafto%20Different%20Paths.pdf
Tronolone, C. (n.d.). *Noam Chomsky/universal grammar.* Retrieved from http://www.rit.edu/cla/philosophy/quine/universal_grammar.html
Vanderbilt Kennedy Center. (n.d.). *Brain development and plasticity.* Retrieved from http://vkc.mc.vanderbilt.edu/topics/brain-development-and-plasticity
Vygotsky, L. S. (1962). *Thought and language* (E. Hanfman & G. Vakar, Eds. & Trans.). Cambridge, MA: MIT Press.
Vygotsky, L. S. (1986). *Thought and language* (A. Kozulin, Ed. & Trans.). Cambridge, MA: MIT Press. (Original work published 1962)
Werker, J. F., & Curtin, S. (2005). PRIMIR: A developmental framework of infant speech processing. *Language Learning and Development 1*(2), 200.

3

Measuring Outcome

Evidence-Based Practice

How is clinical effort traditionally documented?

To effectively assess and treat dyslexia, clinicians develop a treatment plan that matches the client's diagnostic profile, apply treatment strategies, and conduct subjective and objective testing at intervals to assess progress. Decades of research have attempted to draw conclusions about best assessment tools, best intervention tools, best treatment methodology, and best measures of outcome to determine effectiveness in reading remediation.

How are educators addressing instructional effectiveness for reading?

For the past few decades, particular focus has been given to quality assurance efforts, referred to as evidence-based practice (EBP). Most EBP research for written language is still emerging in terms of what is best. The research seems to be coming from two directions that sometimes merge. One line of EBP research addresses educational management of reading, examining methods of identifying reading levels, teaching reading, monitoring reading development, and providing different tiers of intervention for children lagging in reading acquisition (Gersten et al., 2009, p. 6).

How is ASHA addressing clinical effectiveness of reading intervention?

The other line of EBP research for speech-language pathology involvement in reading has been addressed by the American Speech-Language-Hearing Association's National Center for Evidence-Based Practice in Communication Disorders (Coleman, Venediktov, Troia, & Wang, 2013). In their review of research addressing literacy in children with developmental language disorders, they state that, to draw conclusions and offer guidelines for best practice, researchers will need to clarify what reading skills are being addressed, delineate the severity of the reading disability, and establish operational definitions of the population being researched, treatments, and outcomes. The center recommends that writing intervention research be inclusive to participants with spoken and/or written language-learning difficulties and exclusive to certain other conditions. Practice suggestions are that a variety of reading intervention categories have a positive impact on language-based learning difficulties such as developmental language delay, especially interventions using the category referred to as synthetic activities, or methodologies based on word decoding at a phonemic level.

How is dyslexia designated in communication disorder research?

Articles cited in the center for EBP report classify reading difficulty using several designations, for instance, dyslexia, developmental aphasia, developmental language disorder with dyslexia, developmental language disorder without dyslexia, specific learning disability, language-learning disability, or specific language disability. Such a wide range of disorder classifications raises many questions about the homogeneity of the population being studied and any conclusions drawn from such disparate terminology. The studies also vary in terms of professional disciplines involved in treatment, the basic nature of programs (educational or clinical), diagnostic classifications, intervention programs, measurement tools, and delivery settings, which may account for the statistically marginal outcomes, with hidden successful strategies probably buried in the difficulty controlling variables. The culling process ahead will be arduous in trying to identify evidence for best practices. It might be helpful for future research to use the National Institutes of Health's definition of dyslexia, that is, unexpected difficulty learning to read, as a starting point and limiting research to oral and written language diagnostic and treatment procedures conducted by speech-language pathologists (SLPs).

How does the SLP proceed with diagnosis and treatment of dyslexia?

In the meantime, the SLP still has the capability of assessing concerns about reading difficulties, selecting diagnostic tools and administering them, arriving at diagnoses, developing a plan of treatment based on findings, and beginning and adjusting treatment depending on client response so that if the strategy that usually works does not, a change is made, sensing patterns of change in client responses, administering objective and subjective measures, providing statements of what changed and how much, and recommending next steps. This may not be an exactly replicable process, but no client is exactly replicable either. Clinicians can be assured that research will continue to guide understanding of communication disorders, and they will welcome input concerning effective treatment tools, methods, and strategies that will eventually be validated in clinical practice.

Outcomes of Phonology Focus in Early Years

How does a child show readiness for beginning reading?

Blachman (1991), in *Getting Ready to Read*, cites an important study comparing two groups of Danish kindergarteners, one with structured, systematic phonologic perception activities over the year with no reference to letters and the control group with unstructured, spontaneous reference to phonologic awareness during play activities. The experimental group outperformed the control group in spelling at the end of first grade and outperformed the control group in spelling and reading at the end of second grade (Lundberg, Frost, & Peterson, 1988). Liberman, Shankweiler, and Liberman (1991, pp. 11–12) note that the only way transcription of words to letters can make sense to a beginning reader is for the child to understand that the transcription has the same number and sequence of units as the spoken word. It requires an intensive phonologic perception training period for a child to have a firm grasp of phonemes and phoneme patterns in words. One year of kindergarten exposure to phonologic perception activities without reference to letters seems to be an excellent start to written language acquisition.

What is the difference between discovering and being told about patterns?

At some point, children will discover that they can expect a phoneme to be associated with one letter more often than others, but they will also discover that they must consider other options as well, maybe six or eight or even more, and then store in memory the correct letter choice for that word. Telling children about patterns and leading them to discover patterns are two entirely different processes. Leading children to understand creates knowledge as they experience varied input, discern the differences, find the patterns, and experience discovery. Discovery makes the rules theirs. In written language, the process begins with listening, is confirmed with listening, is stored through listening, and can be reevaluated through listening, even though the process involves visual letters.

Does NAR apply to adults?

The Neurolinguistic Approach to Reading (NAR) program is often described with reference to children because, fortunately, we now recognize the condition of dyslexia early enough that intervention typically begins in childhood, albeit later than would be ideal. Many adults, however, also have dyslexia to varying degrees and have not received remedial treatment for their inefficient reading for various reasons. One of the important features of NAR is that it can be used with preschool, elementary, middle, and high school children; college students; and adults. The stages of this program are relatively few and relatively simple, but consistency of effort over sufficient time to gain mastery is important.

NAR Outcomes

How has NAR effectiveness been measured?

Most indicators of the effectiveness of NAR are based on subjective, clinical judgment. Hundreds of children have been exposed to systematic application of NAR principles, strategies, and activities at The Kamara Center (TKC) over the years. Objective testing has overwhelmingly documented statistically significant progress. Reading and writing scores have dependably improved, sometimes gradually over years, other times dramatically over 1 year. Clinicians have been confident that NAR is the critical factor that accounts for written language gains. Despite clinical confidence that the program is effective, it has not been possible to isolate NAR as *the* factor that accounts for client gains.

What variables need consideration in effectiveness assessment?

Perhaps the most limiting factor to verifying effectiveness is that NAR is not the only treatment being administered to children because their varied profiles require varied treatments. Typically, dyslexia is not a stand-alone communication disorder, and other strategies and programs must be applied to different children who exhibit other conditions in addition to dyslexia such as abnormal auditory processing, dysnomia, dysphasia, dyspraxia, short-term memory deficit, working memory deficit, articulation disorder, disfluency, dysarthria—whatever other conditions SLPs treat. Still, clinicians can sort responses to each regimen's effectiveness by how the child responds to a series of tasks. Granted, gains in one area can boost gains in another, but even that can be analyzed clinically. Occasionally, posttreatment testing results in unanticipated findings, more often positively, occasionally negatively, but usually clinical judgment gathered from client responses to activities coincides with posttreatment test findings. Perhaps the best way to preserve an on-course treatment plan with accurate assessment of progress is to maintain a reasonable caseload that allows clinical judgment to guide treatment with objective testing confirming outcome. This may not serve as a national best practice standard, but it is reassuring to the child's parent or adult and the clinician delivering the service.

What makes human behavioral research difficult?

Stripping variables to allow parallel comparisons of different dyslexia treatment outcomes is extremely difficult and brings into question the applicability of findings to an individual, real, and complex child. Research may support broad generalizations such as the findings that increasing phonologic perception improves reading, but even these studies often have marginal results, leaving questions about the studies' ability to control for children's clinical profiles, clinician variables, adherence to the treatment protocol, and basic distinctions made between phonics and phonology in the program design. Even research on discrete, step-by-step computer programs such as Fast ForWord have reported difficulty controlling variables such as protocol adherence. Perhaps clinical judgment functions like the language acquisition device, sorting through all the variables to find the relevant patterns!

To what is clinical outcome ultimately matched?

These points are not meant to disparage research but to alert that it is the blue ribbon panel of experts that establishes the benchmark for effectiveness. Researchers are simply trying to ferret out what factors the blue ribbon panel is using so that, hopefully, less-experienced clinicians can consider and implement the panel's view.

References

Blachman, B. A. (1991). *Getting ready to read: Learning how to print maps to speech.* Timonium, MD: York Press.

Coleman, J. J., Venediktov, R. A., Troia, G. A., & Wang, B. P. (2013, July). *Impact of literacy on achievement outcomes of children with developmental language disorders: A systematic review.* Retrieved from http://www.asha.org/uploadedFiles/EBSR-Impact-of-Literacy-Intervention.pdf

Gersten, R., Compton, D., Connor, C. M., Dimino, J., Santoro, L., Linan-Thompson, S., & Tilly, W. D. (2009, February). *Assisting students struggling with reading: Response to intervention (Rtl) and multi-tier intervention in the primary grades* (NCEE

2009-4045). Washington, DC: Institute of Education Sciences, U.S. Department of Education. Retrieved from http://ies.ed.gov/ncee/wwc/pdf/practice_guides/rti_reading_pg_021809.pdf

Liberman, I. Y., Shankweiler, D., & Liberman, A. M. (1989). *The alphabetic principle and learning to read* (pp. 1–33). Ann Arbor: University of Michigan Press.

Lundberg, I., Frost, J., & Peterson, O. (1988). Effects of extensive program for stimulating phonological awareness in preschool children. *Reading Research Quarterly, 23*, 263–284.

4

Other Spelling/Reading Programs

Phonics Versus Whole Language

Is NAR incompatible with whole language or phonics?

One advantage of the developmental approach to reading and writing used by the Neurolinguistic Approach to Reading (NAR) is that it bypasses the controversy between phonics and whole language. The approach NAR uses sides with neither phonics nor whole language. Rather, it focuses on the underlying linguistic deficits that impede reading and writing acquisition. Once these deficits have been ameliorated through the types of activities described in NAR, whole language and some phonics rules will be welcome approaches to reading and writing. These approaches have potential to reinforce coding fluency and build appreciation for a set of communication events that can culminate in the production of a creative, child-written document that the child can read.

What does struggle sounding out words indicate?

Premature presentation of either phonics or whole-language instruction can result in significant frustration in learning. For instance, the child struggling to use a phonics strategy of sounding out an unfamiliar word, saying, "b-b-bee-bee-been-bench-ch-chen-ben" for *bench,* is really at a prephonics stage. The child needs to go far back in developmental learning and listen for sounds in words in a variety of activities until confidence is obtained about what sounds are in the words of known vocabulary and not fail to recognize that the word had already been pronounced correctly but was not recognized because of concentration on the letters.

When would a child be ready for "whole language"?

Likewise, the child with dyslexia who sits in terror with excited classmates as they write a news release for a play they just wrote and performed about their visit to the zoo, praying that the teacher will not single anyone out to spell anything, is not ready to embrace the wonderful world of whole language. That fearful child must first be able to grasp and apply the *alphabetic principle,* that is, words are made up of phonemes that are represented with letters and, to do this, must have an adequate array of auditory perceptual skills used in the phonology rule system of language.

Even maximum motivation and interest for a reading/writing project cannot prepare the child to code sounds and letters. However, once the coding process is reasonable, not perfect, the creativity of whole-language strategies should afford the child expression of ideas, feelings, and opinions in a uniquely creative manner, a worthy goal.

How widely held is the concept of alphabetic principle?

The term *alphabetic principle* is not new. In the 1960s, Heilman used the term in alerting teachers to consider phonologic perception as a basic skill in the coding process (Heilman, 1976, pp. 1–19). The term became more frequently used in later years, for instance, in the seminal publication, "The Alphabetic Principle and Learning to Read" (Liberman, Shankweiler, & Liberman, 1989), distributed by the National Institute of Child Health and Human Development. Many other research articles have used this term for the past several decades, which raises questions regarding minimal implementation of the principle to individual children in assessing their written language progression.

Does dyslexia mean lifelong avoidance of reading and writing?

An interesting feature of many children with dyslexia is that, once they can manage coding, much stronger semantic skills are typically released, and they generate imaginative, rich, and sometimes profound narrative and expository products. Many even become voracious readers. Comprehension and expression of meaning were never their problems; the coding struggle was simply so frustrating and energy consuming that they had no cognitive energy left for comprehending what they read and wrote, so they simply rejected reading and writing. Some children hold on to negative feelings about reading and writing, even after they have become proficient, because they cannot forget the embarrassment.

Lindamood

What did Lindamood code with colored blocks?

Some speech-language pathologists (SLPs) have found inventive ways to address dyslexia. Speech-language pathologist Pat Lindamood was one of the early proponents of a remedial program that dealt with the basic perceptual phonologic deficit that she found in her clients with dyslexia. Through her own intuitiveness, she discovered that using colored blocks to spell words allowed her to detect what phonemes were not heard, misheard, added, put in the wrong position, or substituted for the ones needed. It allowed her to target the error patterns with supplemental exercises to help clients perceive more precisely the phonemes to be coded with letters in words, which led her clients with dyslexia to more successful reading. She developed a training program called Auditory Discrimination in Depth (C. H. Lindamood & Lindamood, 1975), later called Lindamood Phoneme Sequencing (LiPS) (P. Lindamood & Lindamood, 1998), which included distinctive feature classification, multisensory discrimination training, nonsense syllable block coding, some consideration of letter coding using letter tiles, and other exercises.

What role has Pat Lindamood played in the treatment of dyslexia?

Many of Lindamood's principles can be found in NAR. Often clinicians use nonletter symbols to represent phonemes, but her classifications and block coding provide a broader visual reference for what children hear as they consider the repertoire of Standard American English (SAE) phonemes. NAR expands the range of distinctive features, characterizes and addresses several features differently, avoids any reference to letters until later in the remedial process, does not offer routinized exercises administered by non-SLPs, and offers a systematic letter coding process for all 40+ phonemes of English using graded spelling lists. Many other differences are found in NAR, but much gratitude is extended to the late Pat Lindamood, who was an expert clinician, a nationally recognized leader in reading research, and an inspiration for the diligence she showed in helping individuals understand the role of phonologic perception in dyslexia. Many SLPs use the LiPS kit for discrimination training and find this a useful resource in working with dyslexia.

Phono-Graphix

How does P-Gx approach reading?

NAR principles can be found in resources introduced by other professionals. As the role of phonology in reading became better understood, many professionals interested in reading realized that a more natural approach to reading was to code phonemes with letters (spell) and then read the words. If children could map what they heard with logical letter spellings, it was felt that they could more easily reverse the process for reading. Shankweiler and Lundquist (1993), in a report for Haskins Laboratories, point out that the processes for spelling and reading are not exactly reversed but that the developmental order of spelling before reading is a more natural process. Phono-Graphix (P-Gx) is one of the programs that uses the concept of spelling sounds. It selects approximately three fourths of the phonemes of English and presents some of the spelling options for each of these phonemes, using a set of early elementary words to exemplify the spelling options. After the child finds the right spelling for a word, the words are grouped with others that share the same letter patterns. Other exercises identify same letters that make different sounds.

How does P-Gx differ from NAR?

Using lists of words to be spelled is one of the concepts NAR uses to target different phonemes; however, the NAR lists are grade based, include many more words that are spelled and read by children at those grades, and cover all 40+ phonemes of English. P-Gx's simplicity is appreciated by parents and provides homeschooling teachers a helpful strategy to correlate spelling with reading. An SLP using P-Gx would notice that several phonemes are not included for coding and that the range of spelling options for the ones cited does not include many of the English spelling options children face in grade-level material. P-Gx offers a greatly simplified concept of phoneme-grapheme spelling. For instance, it does not include the highly problematic spelling options for [r] and [ɝ]. It offers no classes or sensory features of phonemes that could serve as recall hooks, and, most important, it does not address the pre-letter-coding phonologic perception skills that block spelling and reading acquisition in many children, especially those with dyslexia. P-Gx presumes a certain degree of reading capability, which is understandable, considering the natural inclination of children to attempt and for adults to encourage word-gestalt, sight reading when young children have inadequate concepts of phoneme-grapheme match. P-Gx attempts to clarify the match for words already word-called correctly and teaches new words by emphasizing the letter pattern of the targeted phoneme. For some children, the approach is sufficient, and in early stages of spelling remediation, after phonologic perception training, SLPs find it a helpful tool for parents to use as a supplement to therapy activities, and some use P-Gx as an approach to reading in the classroom.

Phonics

Do OGB programs teach how to pronounce letters or how to spell sounds?

The commonly implemented approach to reading in schools is phonics, and the construct for that approach is found in Orton-Gillingham-based (OGB) programs. Teacher commitment to these programs tends to be resolute. The rules for reading (how to pronounce written words) are presented as a set of instructions using terminology that requires *if-then* reasoning for children to understand, memorize, and apply to written words. It requires short-term retention of rules presented during instruction, varying degrees of reinforcement depending on the child, accurate storage, and recall of rules from long-term stored memory for application when actually reading and spelling. In other words, the OGB rules are another translation of written language graphemes, noting that graphemes are already a translation of oral language phonemes. These programs tend to be highly structured with numbered lessons and instructional levels and meet the National Reading Panel's requirement for being explicit and systematic. Again, many homeschooling parents find OGB programs helpful because they offer a step-by-step process, which parents can modify because homeschooling can be more readily individualized.

How successful have OGB programs been?

Despite widespread use and a great deal of research, the evidence for reading success using OGB programs is not evident. *Best Evidence Encyclopedia* (Johns Hopkins University, School of Education, n.d.), *What Works Clearinghouse* (Institute of Education Sciences, U. S. Department of Education, 2010), and the *Journal of Special Education* (Ritchey & Goeke, 2006) all cite the lack of conclusive evidence for the effectiveness of OGB programs. Considering the widespread use of OGB programs, their lack of evidence for success, and the continuing reading literacy problem in the United States, providers of remedial reading services may need to reconsider the National Reading Panel's first recommendation for reading success, phonologic perception training.

What skills help children succeed with OGB instruction?

This author's experience supervising many SLPs for many years in a school for children with learning disabilities allowed observation of the difficulties language-learning impaired children experience in classrooms using an OGB approach to reading. The author and other clinicians found many of these students showing significant frustration during classroom instruction and later in their comments about reading. As children with dyslexia, they tended to have weak memory as well as weak phoneme perception, so they could not cite or explain rules that applied. Much of the invented rule terminology had no references in their lives, so they had no clues for meaning and had to memorize, but not really know how to apply, words such as *digraph, vowel diphthong, open syllable, macron, breve, r-controlled syllable*, and *VCe long*. They could not cope with inconsistent rules that did not apply when they tried to use them. SLPs noted that some rules were phonetically inaccurate, and in critical areas. Some children with stronger short-term memory and long-term recall skills who were more agile in *if-then* reasoning could skip over inconsistencies and irregularities and apply the rules during the split-second decoding process in reading and longer, but still short, encoding process in spelling. They, rightfully, took pride in their ability explain the rules for how words should be pronounced. Their brains automatically sidestepped rules that did not fit and filled in the coding process using other strategies such as context to cue words, much stronger visual whole-word recall, predictability of word order, syntax knowledge as a native speaker, and their intelligence. These less-involved children or children with compensatory resources could benefit from the systematic structure of the OGB program. But children with more complex and severe dyslexia required much more support in speech-language therapy to master reading.

References

Heilman, A. W. (1976). *Phonics in proper perspective* (3rd ed.). Columbus, OH: Charles E. Merrill.

Johns Hopkins University, School of Education. (n.d.). *Best Evidence Encyclopedia: Orton-Gillingham Approach.* Retrieved from: http://www.bestevidence.org/overviews/O/Orton-Gillingham-Approach.htm

Liberman, I. Y., Shankweiler, D., & Liberman, A. M. (1989). *The alphabetic principle and learning to read* (pp. 1–33). Ann Arbor: University of Michigan Press.

Lindamood, C. H., & Lindamood, P. C. (1975). *Auditory discrimination in depth* (Rev. ed.). Austin, TX: Pro-Ed.

Lindamood, P., & Lindamood, P. (1998). *The Lindamood phoneme sequencing program for reading, spelling, and speech* (3rd ed.). Austin, TX: Pro-Ed.

Ritchey, K. D., & Goeke, J. L. (2006, November). Orton-Gillingham and Orton-Gillingham-based reading instruction: A review of the literature. *Journal of Special Education, 40*(3), 171–183.

Shankweiler, D., & Lundquist, E. (1993). *On the relations between learning to spell and learning to read.* New Haven, CT: Haskins Laboratories.

What Works Clearinghouse, Institute of Education Sciences, U. S. Department of Education. (2010, July). *Orton-Gillingham-based strategies (unbranded).* Retrieved from: http://ies.ed.gov/ncee/wwc/pdf/intervention_reports/wwc_orton gill_070110.pdf

5

Overview of NAR

Four Stages of NAR

What is the difference between phonics and phonologic perception?

Before presenting the stages, a basic concept that requires consideration is the difference between phonics and phonology as applied to phonologic perception, including the role of the alphabet and the sounds associated with letters. This is a basic distinction that needs to be applied at every level of the Neurolinguistic Approach to Reading (NAR): Phonics begins with letters (or graphemes); phonologic perception begins with sounds (or phonemes). When adults tell children to look at written words and sound them out, they are using a phonics strategy. Phonics begins with vision (i.e., looking). By contrast, when adults ask children what sound is at the end of a word, they are engaging phonologic perception. Phonologic perception training begins with audition (i.e., listening). Some reading instructors merge the two strategies and say that each includes the other, but they are fundamentally different. Again, phonics references letters; phonologic perception references phonemes.

What is phonetics?

A third term that also requires consideration is phonetics, which is a phoneme coding system that has one-to-one correspondence between the phoneme and the visual symbol representing that phoneme. Unlike Standard American English (SAE) spelling, no multiple use of symbols or phonemes occurs. The International Phonetic Alphabet codes thousands of phonemes in languages around the world, using visual symbols, but, of course, NAR targets SAE phonemes. It is interesting, however, to note that many speech production near misses in spontaneous speech, oral reading, and spelling involve features of non-English phonemes, and those errors often give clues to speech-language pathologists (SLPs) who routinely use phonetics to record actual versus target phoneme production. For instance, a child may lip-round a lip-spread vowel, resulting in a German-sounding phoneme, and that child might also spell *said* as *sewd* in trying to represent a lip-rounding element that actually is not in the word but is misheard in silent rehearsal. A pattern of lip-rounding/spreading errors would suggest need for labial feature clarification for vowels and consonants. Or if dentalization errors for glide /w/ or /r/ were part of an earlier or current articulation profile, with spelling errors such as *vent* for *went* and *vabbit* for *rabbit*, treatment would target differentiation of labial elements.

When does phoneme perception begin?

The auditory process of phoneme and word recognition begins oral language acquisition at the earliest stages of life. Any misperceptions not clarified at these early stages of coding sounds as phonemes have potential to be carried throughout the entire oral and written language-learning sequence. Some clues of breakdown are very apparent, for instance, when a child says *baf* for *bath* and writes *elethunt* for *elephant*, reflecting perceptual merger of two phonemes into one. Detectable errors such as these are usually the tip of the iceberg in terms of a confused phonologic rule system, and, by the time this type of error can be identified, the child is often well along in the reading and writing instructional process and has carried this confusion for a long time. That child needs to go back to an earlier stage of phonologic perception training, not just for speaking, reading, and spelling but for listening as well. A strong phonologic perception training program will help rectify confusions and enable the child to process and produce phoneme patterns fluently. It should also be considered that phonologic confusion may be the cause of, or at least part of, other language vulnerabilities found in the child with dyslexia. Imagine applying morphologic and semantic rules to ambiguously perceived phoneme strings.

Stage I: The Alphabet

Should the sounds of letters be taught at alphabet training level with NAR?

The first stage of the NAR program involves teaching alphabet knowledge as a separate skill to be applied to phonemes at a later stage. It may seem odd to start with the alphabet because NAR's primary focus is on clarifying phoneme patterns. However, the alphabet must be dealt with realistically. It will be used eventually, and it represents a heavy memory task for the preschooler. It is better for the child to initially learn the alphabet unencumbered by phoneme-grapheme coding—or miscoding. Stage I is often moot for children beyond second grade because they have already learned the alphabet, but it is still important to assess alphabet knowledge as a memorized and applied system, keeping in mind that many bright high schoolers cannot alphabetize.

Stage II: The Mouth

Why is the mouth important to reading?

Critical listening skills in relation to mouth position and movement are established at this stage. Intensive auditory training is used for recognition of the 40+ phonemes of SAE, using tactile, kinesthetic, and visual cues that help clarify weak auditory skills. Phoneticians have different opinions about a few phoneme patterns, for instance, whether certain allophonic variations constitute a unique phoneme or are regional variations not widespread enough to be considered an SAE phoneme. Most research cites the number of SAE phonemes as 40 and possibly a few more. Phonemes are grouped according to common features such as *popping*, *voicing*, or *hissing* characteristics to facilitate memory. Discussion and sensory cues facilitate the sensory analysis. The child learns articulator placement and movement patterns for different sounds, detects vibration of the vocal folds, and determines whether sounds exit the nose or mouth, all in relation to the acoustic outcomes of these placements and processes. Even though children have experienced their auditory environment their entire lives, typically, they have limited awareness of exactly what the mouth does to differentiate phonemes. One of the first tasks is to sensitize the mouth to features that result in different acoustic qualities so that there is better recognition of all 40+ phonemes of SAE before beginning to code them with letters.

Stage III: Nonletter Spelling (Color Coding)

What do colored items represent in NAR?

Auditory discrimination is strengthened and tested in this next stage. Phoneme patterns are coded with colored objects: small plastic toys, blocks, flat marbles, or other objects that allow discrete representation of approximately five to seven (or more for older children) phoneme elements in a word to demonstrate how many sounds are heard, what their sequence is, and whether any sounds occur more than once in the sequence. Vocabulary/spelling lists found in the grade-level spelling sets that are appropriate for the age, grade, cognitive ability, and reading level of the child are used as the auditory stimuli, which the child codes with colored blocks. The lists are categorized to follow the groupings used to teach about the mouth and how it produces sounds. Nonword lists can be found in Appendix D to carry out nonletter coding, and lists in other kits and manuals can be used for color coding by applying the same principles.

Stage IV: Letter Spelling

How are words pronounced for spelling?

In the final stage, for the first time, the possible ways to spell the 40+ phonemes of SAE are introduced, matching phonemes and graphemes auditorily *and* visually, using the same lists of words and nonwords spoken in the block-coding stage. All consonants, vowels, diphonemes, and glides and a few frequently occurring phoneme combinations are presented in the lists. Each grade-level set has lists of words that represent the ways each targeted phoneme is spelled for a vocabulary at that grade level. NAR provides a comprehensive representation of the SAE phoneme repertoire, with SAE pronunciations used to categorize lists. Obviously, regional and cultural dialects involve different pronunciations of certain words, which will need to be calibrated to SAE in the same manner that spelling pronunciation in different dialects and cultures has always been calibrated to SAE pronunciation. When explanation of an implicated phonologic rule would be too involved, the rule is bypassed. These rules usually involve the physics of coarticulation that result in inconsistently spelled phonemes; for instance, the [t] is spelled in *pants* but not in *dance*. The confusion can be noted and the young child complimented for good listening in hearing the [t] in *dance*, but it is not necessary to explain the physics of why it occurred. The goal of the program is not to make children phoneticians; rather, it is to give them reasonable auditory references for spelling and reading. They have access to many distinctions that they may not need to apply, but a broad spectrum of patterns is available for use as needed.

Adults

Do adults recall the skills they gained that allowed them to learn to read?

Well-meaning teachers, supervisors, parents, television programmers, and promoters of commercial reading products generally do not take into consideration typical and atypical oral and written language development. Many adults know and can talk about advanced-level orthographic rules for spelling, but they have forgotten or never needed to think about the underlying oral language rules they acquired as children upon which many of their conclusions about print rules are based. They cannot remember what it was like to be a child learning the rules of language, for instance, those rules learned during the critical infant stages of matching mouth movements to speech sounds; they have no recall of those ages. They cannot even remember what it was like for them to sort the rules that were presented in their early elementary language arts classes, except perhaps some vague recollection of success or failure.

What is the risk of spelling based on complicated and inconsistent letter rules that need to be recalled?

Although mature adult speakers and writers have the capacity to look back on the language-learning process and discover some of the rules that their language acquisition device (LADs) automatically employed in early years of life to sort the enormous amount of data they received about phonemes and letters in later years, retrospective rule citing for spelling is still somewhat of a conjecture process. A convoluted and inconsistent rule-citing process only adds another layer to the spelling/reading task. An older, successful reader/writer may want to refine some spelling patterns such as the doubling of /r/ depending on changed syllable stress with addition of a suffix, but even then, the rule may be too complex to be readily recalled for application, and the person might be better off to just practice spelling the word many times or conceptually pairing it with a known word that is spelled with the same pattern.

Can written language be learned automatically through exposure to print?

Still, adults tend to teach reading to children through letter patterns because adults have succeeded, to varying degrees, to decipher letters. They think in reverse to arrive at pronunciation of words. Adults no longer need the foundation of listening to and saying sounds in words to anchor each step of the spelling and reading process as children just learning to read and write do. This is unfortunate because, unlike oral language, written language must be taught. Just being exposed to written documents will not yield reading, writing, and spelling capability, except for a very small number of children, and a significant number of these children who display exceptionally fluid preschool reading still show reading comprehension difficulty. Written language is unlike oral language in which simple exposure to the language is sufficient.

What is problematic about phonemic awareness exercises as implemented?

Research remains our guide to understanding language learning, and research has clarified that children who know the sounds in words have a much greater reading success rate than do children who do not perceive phoneme patterns in words. As a result, kindergarten and first-grade curriculums now include phonemic awareness exercises. In reviewing textbooks for teacher preparation and observing classrooms, however, it is clear that the scope and intensity of the exercises are not sufficient. The activities are often phonics exercises, not phonologic perception exercises, sometimes presuming that children with dyslexia can look at letters and readily match phonemes to them. Some exercises require the child to completely bypass consideration of auditory patterns in words, instead saying what words the letter strings represent, which encourages sight reading. As one very bright first-grade child with hyperlexia explained after being praised by her teacher for her reading, "I can say the words, but I'm not really reading." Although her teacher reassured her that she read well, saying that she was silly, the child was actually right. She had memorized letter patterns for pronunciation, but they did not yield meaning.

At what age do children develop memory and reasoning skills necessary to apply orthographic rules to reading?

Educators at conferences, lectures, and various discussion forums who describe phonics instructional programs typically present a crystallized set of orthographic reading and writing rules that they have come to accept as what the child learning to read should be taught. They readily recite these rules, which listeners often verify as what they also know and have used. Some educators have written reading instruction manuals based on this set of rules, developed workbooks, conducted workshops, and built national organizations with local chapters. Similar product lines have been developed by other educators who espouse the same approach to reading instruction as this new thought continues to be introduced to school districts across the nation. The instructional force is formidable. Yet research shows that the memory and reasoning skills needed to grasp the orthographic rules for reading and spelling cited in these programs typically do not emerge until late- or postreading instruction ages (McGuinness & McGuinness, 1998), and the child with dyslexia is even more disadvantaged than classmates because of the tendency to have weaker memory skills compared with age peers.

What does speech-language pathology need to do to facilitate spelling/reading?

Students preparing to be teachers may be increasingly aware of phonemic awareness and use textbooks that describe activities to stimulate consideration of phonologic features of words. However, even these textbooks make it clear that speech-language pathology has not played an effective role in helping educators understand the oral and written language skills necessary for the process of spelling/reading and the effects that various language, memory, and processing deficits have on that process.

What obstacle do children with dyslexia face with "dumbed-down" spelling patterns?

Obviously, letters eventually will have an important role in spelling, reading, and writing, but reading needs to begin with clear understanding of the phoneme patterns in words. A child must hear—that is, perceive what is being coded—or the process becomes one of rote memorizing and guessing, both subject to breakdown. Letter matching to sounds will come later, but first the sounds that comprise words must be accurately perceived, and when letters are introduced, care must be given to the reality and absurdity of SAE spelling patterns. The dyslexia problem is not solved by "dumbing down" spelling patterns presented to children, for instance, by using a text that repeats several */-et/ family* words (e.g., *pet, set, met, wet*). They have trouble jumping from contrived texts such as these to real texts that spell [ɛ] other ways such as *said, read,* and *any,* still not knowing how to match sounds to letters and becoming further discouraged.

What happens when adults cite rules the child does not understand?

All this discussion of the gap between adult and child understanding translates to a principle of paramount importance: An adult working with children who are struggling to read, even if the adult has had similar struggles as a child, should calibrate instruction to the child's level of understanding and the child's zone of proximal development (ZPD), not the adult's. The gap between adult and child understanding accounts for much of the discouragement profile associated with dyslexia. Downplaying how absolute the presumed rules are and showing some humility at the wondrous job the LAD does for children without dyslexia, and even children with dyslexia, can ease the learning pressure.

Why do educational models for teaching reading flip-flop over time?

Taking these factors into consideration, an additional observation might be added to Vygotsky's ZPD: The zone of learning must take into consideration what the child does know and what the child needs to know to advance knowledge, and it must also take into consideration the adult perception of the concept being addressed. Much of the fluctuation in different models of reading instruction over the years has been based on presumed knowledge that actually needed more exploration before it was formulated into the reading curriculum. For instance, a recurring and recent trend to teach infants and toddlers to sight read words in a visual memorization approach to reading has been rejected because reading for meaning is not a memorization process. Reading acquisition is a distinctively human phenomenon with fundamental skills presented to the child during a very narrow window of time in his life. We have much to learn about differences in children's profiles during that window of time with consideration of neurobiologic systems, maturational schedules, instructional options, and best outcomes.

Reference

McGuinness, C., & McGuinness, G. (1998). *Reading reflex: The foolproof Phono-Graphix method for teaching your child to read.* New York, NY: Free Press.

6

Detailed Description of Stage I: Alphabet Knowledge

When is the alphabet introduced to children?

How and when the alphabet is taught is an area of understandable confusion, beginning with alphabet blocks presented to babies long before they have concepts of written language. Obviously, children will need to have access to letter knowledge to eventually code phonemes with graphemes, but letter name knowledge should be acquired as a separate process from phonologic perception training to be applied to written language coding later on. Keeping letter name knowledge separate from phoneme association makes the memory task simpler and more feasible for mastery. Many teachers expect children to enter kindergarten with alphabet familiarity. The Neurolinguistic Approach to Reading (NAR) suggests gaining mastery as soon as possible, allowing for the variable developmental schedules of individual children. Learning the alphabet includes saying the alphabet, pointing to named letters, naming letters, and writing named letters. The need to follow these steps of alphabet mastery will especially apply to children showing delay or difficulty in reading and writing readiness skills. Notice that these steps in alphabet learning include no phonics requirement to say a sound associated with a letter. In fact, children with dyslexia who have started to make these phonics associations often become frustrated when they encounter four or more letter patterns that are regularly associated with a phoneme. These patterns will need to be carefully and systematically applied to the sound-letter coding process but much later on to keep the memory requirements manageable.

What is this called?

Before discussing a progression of steps that is helpful for learning the alphabet, it could be useful to complete the following exercise, which should provide insight to the process a child goes through in learning the alphabet. Write the word or syllable beside each of the following numbers in Table 6–1 to indicate what each of these symbol sets is saying. Use the phonetic chart (Table 7–1), Key Words With IPA Symbols, in Chapter 7 to help you.

Table 6–1. What Is This Called?

1.	ju	10.	dʌbʊlju	19.	si
2.	waɪ	11.	di	20.	keɪ
3.	vi	12.	pi	21.	bi
4.	dʒeɪ	13.	i	22.	oʊ
5.	es	14.	kju	23.	dʒi
6.	eɪ	15.	ɛf	24.	ɛks
7.	ɛl	16.	ar	25.	ɛn
8.	ti	17.	aɪ	26.	zi
9.	eɪtʃ	18.	ɛm		

Why is it important to have solid alphabet knowledge before matching letters to phonemes?

Yes, this is the alphabet. Some of the phonetic symbols may have been confusing, but then children have to cope with /p/, /ɑ/, /b/, /d/, /g/, and /q/, sticks and balls with arbitrary placements, so the task was similar to what the child faces when trying to read words using letter names. For this reason alone—and there are several others—it is important later during coding activities to say letter names minimally, keeping the coding process a silent, visual-symbol representation of the involved phonemes in reading and spelling activities as much as possible. There will obviously be times when both the child and the clinician will need to say a letter name to clarify a point of reference. Virtually all children have some or much letter name knowledge by the time they are identified as needing assistance in learning to read. The goal is to clarify that knowledge and place it in reserve for later use.

Who can help with letter reversals?

It is important to rectify letter reversals or letter formation confusions as much as possible before the letters are used for coding, or the combination of weak visual motor, visual perception, and auditory perception skills can impede progress. An occupational therapist cognizant of the impact these problems have in learning to spell and read is often helpful with these types of letter formation and perception problems. Stick-and-ball letter charts in Appendix A may be helpful for some children with letter reversals.

Why are vowels [i], [ɛ], [aɪ], and [eɪ] placed before and after the target phoneme in letter names?

The vowel sounds placed before and after a given phoneme element in the letter name actually serve a purpose. They provide a context for the targeted phoneme that clarifies the acoustic qualities of the sound as the mouth moves from one sound to the next. The acoustic transition provides a reference that gives more accurate identification of the target phoneme. The transition can be seen on a spectrograph as a brief rising, continuing, or falling tail of energy (representing the frequency, volume, and duration of a phoneme's acoustic energy). The transitions are referred to as loci. Most languages present the phonemes of the language in the context of syllables, that is, letter names, as English does. Speech science research tells us that the transition element between sounds is critical for discrimination, and its brevity makes discrimination difficult for many children with dyslexia. The normal transition time between sounds in words is extremely rapid, measured in milliseconds. Computerized auditory training programs have been devised that lengthen the duration of the transition to provide more opportunity for distinguishing the targeted phoneme (Tallal et al., 1996).

What common mistake do children make in spelling when "thinking about" letter names?

Letter names provide the literate, older child or adult with helpful information about sounds and letters, but letter names can confuse the child learning to spell and read. Children with immature written language skills spell *are* as /r/ because the letter name seems sufficient. They spell *enemy* as /nemy/, *empty* as /*mpty*/, *pea* as /p/, and the list goes on. Most children learn to cope with the distinctions between letter names and phonemes, but many do not, and even if they do, it can require extra thinking. Children and adults often add /*uh*/ just before or after a phoneme, not realizing that they are saying two sounds, for instance, saying [b] as [bʌ], not just [b]. This leads to confusion in some children (e.g., spelling *but* as /bt/), especially children with dyslexia, when sounding out words. Adding unnecessary phonemes complicates the difficult auditory discrimination task the child with dyslexia already faces.

What effect does weak alphabet knowledge have on later phoneme-letter coding?

Many early prespellers and prereaders are presumed to have letter name mastery, but this is not always true. Their not having ready recall of letter forms and names interferes with the more difficult task of sound-letter coding that they will be facing. Remediation will often require reverting to earlier stages of alphabet instruction to shore up alphabet knowledge as much as possible. In many ways, it would have been better for the older child and adult having reading and spelling difficulty not to have learned to think of letter names when spelling and reading, but the time has passed, and they have probably habituated a pattern of saying a letter name silently or out loud as they sound out words in reading and spelling. Unlearning this pattern is difficult, but some children can be conditioned to reverse their thinking of letter names and instead think of phonemes when reading and spelling.

What does needing to sing *The Alphabet Song* to say the alphabet signal?

Learning the alphabet in a certain sequence usually facilitates the process, and there is no stopping *The Alphabet Song*. Children hear it on TV, CDs, DVDs, interactive books and toys, in social gatherings, and with their families. They often enter school proud that they know it, and they ask you to sing it with them "next time." This enthusiasm for learning the alphabet can be cultivated and used to the child's advantage. Dependency on the melody to cue next letter names should be monitored. Children who sing the song straight through compared to ones who need to re-sing all or parts of it many times are at different stages of memory development. Stumbled recitation usually signals that memorization of the sequence is not yet sufficient. Before the phoneme-letter coding task begins, a child should be able to say the alphabet without need to sing all or parts of it for sequence cues. Being able to do this indicates that the process has been learned to the point of automaticity. The importance of its being memorized is not that the sequence will be used very much in actual reading or spelling, although alphabet memorization helps task organization later on, but rather saying the alphabet fluently is an indication of auditory recall of the names for visual symbols. That name will later be dropped in spelling and reading, but the visual pattern will be retained, doubly stored through two senses, auditory and visual. A side benefit is that the task is an auditory memory exercise that facilitates a skill always needing reinforcement.

What are the four alphabet tasks a child should master before coding with letters?

The four alphabet tasks that the child will need to master are (1) reciting the alphabet, (2) pointing to randomly named letters, (3) naming randomly designated letters, and (4) printing randomly named letters. These tasks need to be executed quickly and accurately before coding phonemes with letters and after phonologic perception skills are solid enough to manage letter coding.

How should handwriting and letter formation be addressed?

Lower case forms should be mastered first. Upper case or a mix of upper and lower case forms can be presented if the child can separate and manage the forms in relation to capitalization rules. Since early readers often do not know those rules, keeping lower case the print form prevents random use of capitalization until it can be introduced systematically and accurately. Printing should be emphasized rather than cursive writing because it yields more legibility; in fact, modified print is recommended as a way to improve legibility in *How To Study in College* (Pauk & Owens, 2008, p. 254), with the key legibility feature being lifting the pencil after each letter. Because many children with dyslexia also have handwriting problems, these options should be considered. Many schools across the United States have dropped instruction for cursive writing for a variety of reasons, including the cost and time factors for teaching it and increased use of computer writing. Individuals eventually modify their writing according to factors such as their fine motor coordination and personal style. Some might question the utility of teaching a skill that will automatically morph for the vast majority of people. Being admired for beautiful handwriting is not a priority for most children with dyslexia, but legible writing is.

What references can be used to help children learn to form letters?

At the stage of writing letters, the occupational therapist may want to give advice about fine motor problems for certain children, but most can learn to write letters using the efficient patterns found in *Handwriting Without Tears* (Olsen, 2014) that are designed for kindergarteners to fifth graders. Another source can be found online at http://www.handwritingforkids.com. The latter reference provides free letter and number models that specify the direction and order of strokes for each letter and number, writing exercises, and lined paper appropriate to age. Parents and tutors find resources such as these helpful.

Various tables are available in Appendix A to assist in alphabet learning. Table A–1 provides random lower case letters for pointing and naming. Table A–2 provides alphabet letters without serifs in order. Table A–3 provides upper case letters. Table A–4 provides lower case letters in alphabetical order. Using the Alphabet Mastery Chart in Appendix A (Table A–5) measuring the accuracy of each of the four alphabet skills needed over several trials can be an encouraging reward system for the child learning to override memorization difficulties.

What are some reasons children use upper case inappropriately?

Because *NAR* uses only lower case letters, except for infrequent proper names or sentence starts, lower case forms are stressed in learning to write the alphabet and are required in later spelling with *NAR*. Children who use uppercase forms because of confusion in letter directionality or ball-and-stick relative positions in lower case forms, for instance, writing /B/ because they cannot recall the directionality of /b/ but they can /B/, should be allowed to use upper case forms until they can differentiate the lower case correlates, but the confusion needs to be explicitly addressed. If not, the confused upper/lower case forms tend to habituate into misuse of upper case forms, especially for sentence segmentation. Young children who demonstrate this confusion in learning upper and lower case letter forms often become college students who hide their confusion about capitalization rules in stylized upper-case handwriting that still is unsuccessful in camouflaging their run-on sentences and sentence fragments. The confusion all begins in kindergarten! Usually the combination of short-term, recent, and long-term retention and retrieval skills a child possesses determines the overload point for the number of discrete information units that can be managed. So if both upper and lower case forms can be taught with inclusion of capitalization rules, the child is ultimately better served, but if memory constraints do not allow retention of such a large data base, focus should be on lower case forms, with later focus on capitalization rules presented separately. For children showing directionality problems ball and stick letters, Tables A–6 and A–7, may be helpful, and Table A–8 provides an example of commonly confused digraphs for practice exercises. Any visual confusions can be addressed in this way.

How rigidly should not saying letter names be applied?

Teaching children to say letter names and recognize and write letters using this process will give them the metalinguistic tools they will need to use when talking about spelling patterns later on, but it will not make them use phonics because they are never asked to say what sound the letter makes, only what letter name goes with the letter. Furthermore, saying the letter name will be used sparingly once they begin to spell and read with *NAR*. After coding skills are strong, saying letter names instead of phonemes will probably do no harm and may strengthen the auditory-visual loop once it is established.

What keyboard features need to be considered?

A few points about computer representation of letters should be considered because they may affect letter recognition and reproduction. The computer keyboard usually displays letters in upper case form, and children who are adept at only lower case forms may have difficulty with the upper case representation, especially with upper case /I/ and /L/, and may need some orientation and cueing. Also, most electronically produced fonts have additional marks called serifs that can confuse children. Finally, typed letters /a/, /q/, and /g/ tend to look different than their handwritten manuscript forms. Some children will need to have these features clarified. In general, children tend to have their own patterns of confusions, and it requires adult monitoring to catch these confusions early on so that they do not become habituated. Preschoolers typing on a computer keyboard is an increasingly common phenomenon as information technology permeates everyday life, and parents can be helpful in clarifying these early alphabet typing confusions.

What clues to obstacles can adults capture as children learn the alphabet?

In young children, not only will gaining letter name mastery ease the burden of phoneme-grapheme match later on, but it will also alert to any memory, visual perception, and motor coordination difficulties at an early age. These difficulties generally are not the cause of dyslexia, but they can occur in children with dyslexia, can complicate written language acquisition, and may require separate remediation. Adults or older siblings help a child gain alphabet recognition, naming, and printing skills using discussion of errors and more practice with those letters. The process is sometimes slow, and the child may become frustrated. In that case, routines such as five-minute practice periods twice a day with a small reward—a sticker, verbal praise—should be sufficient to nudge progress. Tracking correct responses on the alphabet charts in Appendix A for each alphabet-learning process will highlight troublesome letters and can be used to motivate the child as the line on the chart grows. If gains are not seen after reasonable effort, however, it may be necessary to allow neurobiologic maturation to take place before starting again. Maturation schedules for skills within a child and compared to other children vary, and too much focus on tasks not yet ready for acquisition should be avoided. Parents of preschoolers typically identify readiness and, if they have concerns, should seek professional consultation.

References

Olsen, J. (2014). Handwriting without tears. Retrieved October 22, 2014, from http://www.hwtears.com/hwt

Pauk, W., & Owens, R. J. Q. (2008). *How to study in college* (9th ed.). Boston, MA: Houghton Mifflin.

Tallal, P., Miller, S. L., Bedi, G., Byma, G., Wang, X., Nagarajan, S. S., . . . Merzenich, M. M. (1996). Language comprehension in language-learning impaired children improved with acoustically modified speech. *Science*, *271*(5245), 81–84.

7

Detailed Description of Stage II: Mouth-Ear Phoneme Perception Training

How does recall cued by an auditory pattern differ from recall with no auditory cue?

Children need a solid base on which they will build written language competency. They need to know how a word sounds because the composite sound of that word is always available as a reference; it does not change. The word can be repronounced slowly to clarify the auditory elements. The sounds in the word are available for confirmation as many times as needed. The constancy of this source, an auditory reference, will later facilitate correct spelling and oral reading. The child can pull information from memory using an auditory sensory reference to stabilize and facilitate the recall.

Distinctive Features

Why is coding phoneme strings important?

Since we do not speak in individual sounds, word segmenting and blending activities should be developed in Stage II and used throughout Stages III and IV as needed to solidify the concept of phoneme strings. For blending activities, individual phonemes or syllables should be said to the child with different time intervals between the sounds, and the child should be asked to blend the phonemes to identify and say the word. For segmenting activities, whole words should be spoken to children, and they should be asked to say each phoneme or syllable in the word in the right order. More of these types of activities are found in Appendix C. Examples of exercises for frequently confused phonemes are provided in Appendix D, Tables D–4 through D–9.

How do multiple senses function in phoneme perception?

Bolstering the phonology rule system using all five senses should be a fun, exploratory stage of training in which the child learns to depend on sensory feedback to give immediate cues to answer questions. Pictures in Appendix I or other resources that exemplify the mouth postures and movements for all Standard American English (SAE) phonemes can be used for reference. Discussion, mirrors, flashlights, graphs, charts, pictures, large-mouth puppets, tongue depressors, coffee stirrers, straws, popsicle sticks, and sterile finger cots can be used to clarify sensory information. The adult and child should compare mouth positions and movements.

What are the characteristics of phonemes called?

Before discussing the phoneme characteristics that children will be presented in Neurolinguistic Approach to Reading (NAR), it is useful to consider the principles of phoneme perception and production that scientists who study phonology within and across languages offer. The characteristics of phonemes that represent the physics, anatomy, and acoustic outcome of articulatory efforts have been referred to as distinctive features and have been delineated in various ways since the early 1930s. Many theories have evolved that consider the rules that the brain uses as a template for understanding and expressing thought, that is, the grammar of a language. That grammar represents the rules that apply to phonology, semantics, syntax, prosody, and pragmatic meaning, which includes the concept that the phoneme is the smallest unit of language. NAR primarily considers the aspects of SAE grammar that involve phonology.

How has distinctive feature theory changed?

Although some would consider earlier phoneme distinctive feature classifications outdated, this perspective is countered by Sandra Miglietta in "Why we need distinctive features" (Miglietta, 2010). Her review of distinctive feature theories gives some perspective to the level of specificity that distinctive features have been and continue to be studied. She noted that early theorists from the late 1930s through the 1970s—Trubetskoy, Stevens, House, Jakobson, Halle, Chomsky, Fant, Lieberman—all renowned researchers in phonology, presented theories on how phonemes are categorized as vowels or consonants, the difference and interplay between articulation and acoustics features, and the role of motor theory in speech perception. Over the next few decades, other concepts were introduced such as hierarchical ordering of phonemes, phonemes as distinctive feature bundles, and feature redundancy. In the past few decades, theories have explored right and left hemisphere brain responses to phonemes, phonemic representation in the auditory cortex of the brain, and time spans of speech units. The newer concepts do not negate original concepts of feature classification; rather, they represent refinements in classification and new theories regarding the processes by which classifications take place.

Why are phoneme maps of world languages important?

Hall and Mielke (2014) also track the history of distinctive feature theory, noting that a major goal continues to be identifying a set of distinctive features that account for all the phonemes observed in the world's languages. Scientists still cannot provide "maps" that fully delineate the phonemes used in the world's thousands of languages. It has not been established what coding functions are mutual, redundant, or exclusive to the coding process for one language compared to another, yet young multiple-language speakers somehow know those maps as they acquire different languages, which reinforces the concept that a universal grammar is operational across languages. White (2003, p. 2) describes a universal grammar as a "genetic blueprint" that determines what forms can and cannot be like, based on requirements of phonologic categories as well as other rule systems such as syntax. She says that a universal grammar allows certain variations across languages while it is based on invariant principles, which she describes as specifically linguistic in nature. In considering a universal grammar, the number of features would need to be expanded from the original 14 binary set to much larger binary sets to classify all languages of the world (Tatham, 1999). Researchers are pursuing these maps, not just for esoteric understanding of a universal grammar but for applications such as speech recognition and synthetic speech production in the era of information technology.

What mutual, theoretical concept is used to treat articulation disorder and dyslexia?

Just as scientists begin with a theory to build understanding of the underpinnings of speech and language, clinicians begin with a theory on which they can provide remediation of speech and language breakdowns. Distinctive features are verifiable through external senses of sight, touch, feel, and hearing. These senses, coupled with cognitive analysis, have traditionally been used with general success to facilitate normal speech production in treatment of articulation disorder. As scientists continue to identify weak phonologic perception as a significant correlate of dyslexia, it is only logical to apply and expand principles used in articulation disorder treatment for treatment of dyslexia.

What problems do speech-to-text software have?

As a linguistic rule system, phonology has many layers of phoneme coding principles that pose intriguing questions about how the human brain processes and produces the sounds of a given language. Universal grammar principles are explored in research that attempts to parse letters of words in text and use the phoneme and morphologic elements within those words to yield text-to-speech translation across languages (Hertz, Younes, & Zinovieva, 1999). Software that translates speech to text has been marketed for the past few decades using phonemes as the coding reference for the text-to-speech programs. Experiences in using many of these programs show that dictation results in errors that can be so problematic to repair that benefit is lost, especially in children whose articulator and resonance structures emit speech as different from adults'. Even the most expensive speech-to-text programs make enough errors that if a person actually spoke the way the speech was translated, that person would be considered to have a serious pathologic condition! But electronic devices are given more allowance for error than humans. The receiver in speech-to-text processing that "hears" speech attempts to execute the same spelling process readers and writers with dyslexia face. The program must analyze the features of the sounds it hears, decide what phonemes they represent, process the phoneme patterns for words they might represent, use the same inferential rules humans use such as word predictability based on context, and then spell the words. The computer often makes the same type of spelling errors children with dyslexia make when they misperceive phonemes in the word to be spelled. Despite limitations, it is important to monitor improvements in speech-to-text research for principles that apply to dyslexia in hopes that, as software improves, more insight to dyslexia may be gained that could be applied to treatment and allow better use of speech-to-text programs as compensatory tools.

What is a distinctive feature bundle?

For purposes of clinical remediation, NAR uses several theories about how the central nervous system (CNS) and central auditory nervous systems (CANS) register phonemes to allow the neurobiologic differentiation of sounds in listening, speaking, reading, and writing. Each of these communication modes can reflect misperception and misapplication of phonologic rules as core issues in oral and written language disorders. These errors provide significant clues as to what is causing breakdowns to occur. A primary concept is that each phoneme is made up of a variety of characteristics. Some phonemes are made with the voice box turned on, some are made with the lips, some are made with the back of the tongue, some are made with the teeth, some pop, some hiss, some resonate in the nose, and so on. Each phoneme has its own unique set of these characteristics that makes it different from every other phoneme in the language. This set is referred to as a distinctive feature bundle. Whether the CNS uses an efficient 6- or 7-feature bundle or a redundant 8- to 14-feature bundle for sorting and identifying phonemes in English is not a critical issue that blocks use of current knowledge about phoneme features to be applied to dyslexia. Articulation therapy has been based on the theory of distinctive feature bundles as applied to articulation disorder by speech-language pathologists (SLPs) for many decades with effective outcomes. NAR applies these same distinctive feature bundle principles to improvement of auditory perception for the purpose of establishing a firm base on which letters can be matched to introduce written language.

How does a clinician gauge the amount of distinctive features analysis needed?

Some children and adults need much study of phoneme bundle features to alter their auditory perception of confused sounds. Others need little, if any. Therefore, remediation must be highly tailored so that unnecessary and lengthy practice does not render the child or adult oversaturated. That being said, it is better to err slightly on the side of giving too much rather than too little practice in analysis of phonologic feature patterns. For the child who seemingly perceived phoneme features automatically and went on to read easily, it may not have been critical to stress this process. However, for children and adults who did not acquire this knowledge automatically, ultimate success in spelling and reading will hinge heavily on how well they master auditory perception of the slight acoustic changes that occur as the articulators make their excursions from one phoneme position to the next.

What is the difference between a vowel and a consonant?

Whether a phoneme becomes a consonant or a vowel is determined by the degree of airway restriction the phoneme receives. The process of phoneme production begins at the point of airflow from the lungs, moving across almost-closed (vibrating) or open vocal folds into the mouth or nasal cavity for vowel or consonant formation, resonating in one of the two chambers, and then being emitted through the mouth or nose. Consonants are produced with greater restriction of airflow than vowels. In other words, the passageway between the surface of the tongue and the top of the mouth is more open for vowels. This relative openness of the oral chamber and the fact that vowels are voiced (i.e., vocal folds vibrate) make vowels louder than consonants. The more restricted the airflow, the more pressure the phoneme has, which affects its volume. Some consonant sounds are louder than others, depending on how much dampening of the sound takes place as a result of the restriction and whether the vocal folds vibrate to produce the consonant. Voiced phonemes are inherently louder sounds, as consonants or vowels.

How can a motor articulation problem affect spelling?

When children display articulation problems due to phonologic misperceptions, they are confused about certain aspects of sound production. Phonologic confusion can also be seen in structure- or motor-based articulation problems because incorrect production of phonemes, no matter the cause, can reset the perceptual *template* for phoneme features. The child's internal template of what constitutes one phoneme versus another is mismatched to the template that others use in distinguishing phonemes. Because phoneme perception precedes phoneme production, which in turn reinforces phoneme perception, any type of articulation weakness, past or present, can adversely affect the perceptual process on which spelling is based, making a child vulnerable in learning to spell and read.

How can distinctive feature misperceptions spread?

Even when children with dyslexia do not have outward indications of speech problems, careful analysis of their auditory perception typically reveals confusion about phonologic processes. For instance, they may say *bathtub* but write /baftub/, indicating residual confusion about the labial and dental features of phonemes, even though they outgrew the articulation confusion. That misperception can lead to confusion about other labial and dental phonemes and spread to other distinctions that ultimately classify the entire phonology rule system.

How are distinctive features coded?

The classification of each phoneme feature is binary, just as information is coded in computer processing. The decisions made about a phoneme feature are based on *plus-minus* (or 0-1) coding and *if-then* reasoning. Each phoneme is coded *plus* or *minus* voiced, *plus* or *minus* dental, *plus* or *minus* labial, *plus* or *minus* nasal, *plus* or *minus* plosive, *plus* or *minus* fricative, and so on. The combination of identified features (*if*) determines whether the phoneme was (*then*) [b] (+voiced, −dental, +labial, −nasal, +plosive, +front, −fricative) or [v] (+voiced, +dental, +labial, −nasal, −plosive, +front, and +fricative). Fricative, plosive, and dental are the features separating these two phonemes. Researchers have modified perspectives on distinctive feature theory over the past several decades to consider that a strict binary decision-making process may not apply to some features. For instance, voicing features of a phoneme can be altered by factors such as surrounding phonemes and syllable stress, so the differentiation of voicing may not be strictly plus or minus. However, for purposes of NAR, the binary system is considered adequate to show the polar aspects of phoneme sorting, allowing the actual language acquisition device (LAD) to execute the finer aspects of continuum sorting.

How can distinctive feature errors indicate severity?

Some phonemes are separated by just one feature; for instance, the only difference between [p] and [b] is voicing, whereas other phonemes are separated by several features, for instance, [θ] and [p] in which the phonemes differ by four features. Just as this applies to articulation errors, spelling errors based on phonologic misperception should be considered in terms of how many distinctive features separate the substituted phoneme from the target phoneme. Errors involving more distinctive feature differences are more serious and may require a thorough review of multiple features, not just the most obvious one. Complete omission of phoneme representation in spelling and reading particularly signals perceptual vulnerability, as is true with articulation errors. Of course, the error would need to be observed frequently enough to determine that it represented a true misperception pattern and not just a random error.

How are all phonemes interrelated?

The matrix of phonemes and features translates to mean that all phonemes are toggled in some way to all other phonemes by virtue of the ± decision made for each feature of each phoneme. This can be visualized in Appendix B, Tables B–1 through B–4, of vowel and consonant features. Table B–1 provides an example of how the three nasal phonemes are distinguished. Table B–2 clarifies glide distinction. Table B–3 provides comparison of all consonant phonemes. Table B–4 provides comparison of all vowel phonemes. All tables are presented as filled-in and blank forms to allow assessment of recognition levels.

If an older child independently completes a targeted part of the chart, the clinician can understand what the child does and does not perceive. The child often says the phoneme many times until the features are conceptualized. Or the clinician can use the blank form to query the child about features that are perceived. Enough wrong ± decisions may show limited ability to perceive features and suggest need for more effort to shift phonologic perception. Some children with severe dyslexia have so many misperceptions that this activity is not productive and will need to be replaced by exercises that consider only a few features at a time. Other children with milder dyslexia may exhibit some difficulty deciphering distinctive features in phonemes but quickly learn to detect the distinctions through phonologic perception activities and go on to achieve a steady base of reference on which to code phonemes with letters.

Is there an order of phoneme mastery across languages?

The developmental order of oral language phoneme mastery from infancy to early and later childhood follows a hierarchical pattern that can be traced using phoneme distinctive features. Individual children offer slight variations as they come to master the phonemes of their language environment, but age-level mastery norms have been presented by many researchers over the years. When the stages of phoneme acquisition across languages are compared, similarities in acquisition patterns are noted, allowing for the fact that different languages have different phonemes (Blache, 1978, p. 113). Studies citing articulation developmental norms are also taken into consideration in written language errors because they can signal phoneme bundle misperception as well as feature misperception detected across letter coding errors.

What kind of spelling errors signal distinctive feature misperception?

Persistent phonologic spelling errors typically indicate that the LAD needs reworking to clarify distinctive features of individual phonemes, phoneme classes, or syllable perceptions. The errors should be monitored for consistency because inconsistency suggests emerging mastery, and the confusion may be resolving spontaneously. Clinical time would be better spent on error patterns that are consistent and represent other phonologic processes. But sometimes even an emerging spontaneous correction needs a nudge because of an excessive age gap or its negative impact on reading. Typical spelling error patterns can be seen in sound substitutions such as *thoup* for /soup/, *elethant* for /elephant/, or *secint* for /second/; syllable or sound duplication such as *ananamle* for /animal/; a tendency to drop whole classes of phonemes such as nasals and glides, as in *stap* for /stamp/ and *wof* for /wolf/; and in-class substitutions such as *sixting* for /sixteen/ and *rin* for /ring/. Children have their own unique profiles of errors, some seemingly random and others showing consistent feature confusion. Some children's errors do not reveal their specific phoneme bundle or feature misperceptions and will need ongoing analysis. The clinician must decide if the errors represent a normal acquisition profile, a delayed but emerging profile, or a deficit profile, keeping in mind that many articulation-delayed children have been later identified as children with dyslexia.

What benefit does understanding the role of distinctive features provide?

Knowing more about phoneme distinctive features helps unravel reading and writing difficulties. It gives the SLP greater insight to areas of vulnerability. It gives the person with dyslexia a scaffold on which to build understanding of what the mouth is doing in relation to the coding of sounds. It gives teachers and parents insight to the minuscule rules that the foundational phonology system uses to build spelling and reading skills. It provides basic understanding of the acoustic consequences of airflow resistance, anatomy, and physics of movement in an extraordinarily small area within the mouth that occur at an extraordinarily rapid rate. These are the kernel rules of oral language on which spelling and reading are built.

The International Phonetic Alphabet (IPA)

Why is the IPA a less confusing coding system for designating phonemes?

Users of NAR may at first find reference to phonetic symbols challenging if they have not had rigorous phonetics training, but phonetic symbols and diacritical marks provide relatively simple, discrete, visual representations of phoneme distinctive features. Each phoneme bundle is captured as a unique visual symbol set. The coding process for training phoneme perception, phoneme production, and letter coding is unambiguous when phonetic symbols are used. The SLP remediating the reading/spelling process should be proficient in phonetics because it is a shorthand, accurate way of representing target versus actual phonemes as perceived, read, or written by the child. Although most adults might assume that they discriminate phonemes well, as many as 30% of adults instructing children with dyslexia have auditory perceptual weaknesses themselves. These vulnerabilities can account for considerable confusion in remediation. Gaining or sharpening phonetic knowledge can provide insight to the coding issues children face, but most adults who work with children with dyslexia should not consider it necessary to become proficient in phonetics or learn all the intricacies of the phonology rule system. Children with dyslexia also do not need to show mastery of the phonology rule system. Fortunately, their LADs still function and do most of the decision making correctly and automatically. They simply need clarification in specific areas, which is what the SLP addresses. One advantage of working with older children and adults is that they have stronger memory mechanisms and can sometimes enjoy learning the phonetic alphabet, which allows an efficient and precise reference vehicle for discussion of phoneme production and perception.

How are letters and phonemes designated?

NAR uses the IPA symbols to efficiently and accurately represent each phoneme. Brackets, [ ], are used to set off phonetic symbols. Brackets are distinguished from slashes, / /, that set off letters. Table 7–1 provides these symbols by groupings as follows.

Table 7–1. Key Words With IPA Symbols

[p] as in *pie*	[θ] as in *think*	[e] as in *egg*
[b] as in *boy*	[ð] as in *that*	[æ] as in *at*
[t] as in *to*	[h] as in *hat*	[ɑ] as in *father*
[d] as in *dog*		[ɔ] as in *or*
[k] as in *cat*	[m] as in *my*	[o] as in *hobo*
[g] as in *go*	[n] as in *no*	[oʊ]* as in *hobo*
[ʔ] as in *kitten*	[ŋ] as in *ring*	[ʊ] as in *book*
		[u] as in *shoe*
[s] as in *so*	[w] as in *we* and *what*	[ɝ] as in *further*
[z] as in *zoo*	[r] as in *run*	[ɚ] as in *further*
[ʃ] as in *she*	[j] as in *yellow*	[ə] as in *above*
[ʒ] as in *measure*	[l] as in *like*	[ʌ] as in *above*
[tʃ] as in *chop*		[aɪ]* as in *eye*
[dʒ] as in *joy*	[i] as in *eat*	[ɔɪ]* as in *boy*
	[ɪ] as in *it*	[aʊ]* as in *out*
[f] as in *fun*	[e] as in *vacation*	[ju]* as in *you*
[v] as in *vine*	[eɪ]* as in *vacation*	

*Diphoneme: two adjacent vowels with one gliding to the other.

By how many features does NAR classify phonemes?

NAR uses several distinctive features for multisensory training that have been identified as important for CNS phoneme identification. Fewer or more features could be used, but these provide a sufficient range of phoneme production characteristics for most children and adults to distinguish sounds. A less redundant set could distinguish all phonemes, but some degree of redundancy can help clarify the phonologic rule system, so more features are included in NAR. The phonetic terminology for each feature is the primary way features are classified, but, depending on the age and inclination of the child or adult, more descriptive or cueing terminology (e.g., *hissing* for *sibilant*) can be substituted. Examples of optional terminology are included in parentheses after each class of phonemes, but other terms can be chosen.

How do children strengthen recognition of features?

Talking about features, demonstrating them by touching the nose or throat, watching in a mirror, and listening for certain acoustic qualities, should set the stage for reinforcement activities such as identifying the common feature of several phonemes. Multisensory facilitation is useful, so the same resources used to explain how sounds are made for speech correction should be used. Children can be quizzed, with errors discussed to clarify their knowledge about phonemes and features, noting how quickly and accurately they make gains. Charts that document performance show progress and can be used to incentivize the child. Turn-taking games with interim questions about features can lighten the burden of the learning task. Some children immediately grasp certain features and accurately identify them in phonemes and words, so they should move on quickly to other features that may be more problematic for them. Other children, especially those with significant phonologic memory problems, require more practice to gain recognition consistently; they often need to review previous exercises to stabilize memory.

Consonant Distinctive Features

How do the grade-level lists categorize phonemes?

Consonant and vowel phoneme classification and phoneme lists found on the grade-level lists can be used for some NAR activities, but the vowel and consonant binary charts in Appendix B provide an explicit reference for analysis of distinctive features for all activities and can be consulted for clarification at any stage in the process. Phoneme properties overlap because each feature is part of a bundle, and other feature bundles (phonemes) may share that same feature. In the following consonant and vowel classifications, each feature will be described, noting phonemes that share the designated distinctive feature. The stylized logo may help conceptualize the key quality of a feature, especially for younger children. Other logos of features are available (e.g., in Lindamood Phoneme Sequencing [LiPS]) or can be devised, or children can draw their own logos, but it is important to represent the full range of features in SAE phonemes if logos are used for reference.

Front (Anterior)

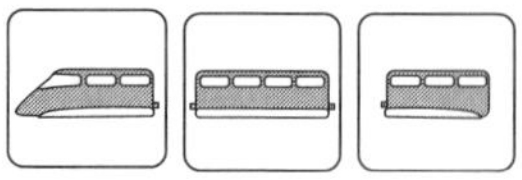

To what does front/ back refer?

This feature refers to phonemes produced in the front part of the mouth where the teeth, lip, or tongue are approximating another articulator to produce a sound. Consonant phonemes for which this approximation takes place in the anterior part of the mouth include [p], [b], [m], [f], [v], [θ], [ð], [t], [d], [n], [s], [z], and [l]. *Back* consonant sounds include consonants [k], [g], [ŋ], [tʃ], and [dʒ].

How can hard-to-see back sounds be identified?

Some of the distinctions between the back and front sounds can be difficult to detect, but differentiating phonemes as front or back can be useful for children having difficulty discriminating phonemes that sound very much alike but are produced in very different places in the mouth, for instance, hearing [d] as [g] and vice versa, a frequently occurring auditory discrimination error for children with central auditory processing difficulty. Hard-of-hearing children also have particular difficulty discriminating these sounds because of the acoustic similarities of the two phonemes and the hidden tongue placements. Deaf children use the Cued Speech hand-signal system to help distinguish the unseen back feature, as well as other mouth features that are difficult to see and for which compromised acoustic information does not allow auditory discrimination. Mirrors, probes, discussion, and contrasting positions can help illuminate this feature.

Labial (Lips)

What qualifies a phoneme to be considered labial?

Five sounds in American English are produced in a way that the lip(s), alone or in combination with the teeth, create the point of airflow resistance. These sounds are [f], [v], [p], [b], and [m]. Often children have difficulty distinguishing the [f] and [v] phonemes from the [θ] and [ð] phonemes, respectively. Sometimes their speech reflects this confusion. Even after speech has matured and these phonemes sound normal, spelling errors often reflect a residual perceptual merger of phonemes, as when the child spells /with/ as *wif* or /brother/ as *brover.* Focusing on the distinction between lip, tongue, and teeth points of contact can clarify the confusion. Lip rounding occurs in certain consonants and glides, including [ʃ], [ʒ], [tʃ], and [dʒ], but the lip involvement does not restrict airflow to the degree that labial phonemes do.

Voiced (Loud, noisy, vibrated)

Are vowels voiced or voiceless?

Phonemes are voiced, or *phonated*, when the vocal folds are approximated, that is, they come so close together that the airstream vibrates the vocal folds and produces tone. Wind musical instruments use this principle by matching the precise amount of air pressure to the instrument opening so that it produces tones. Voiced phonemes include [b], [d], [g], [v], [ð], [z], [ʒ], [dʒ], [m], [n], [ŋ], [j], [l], [r], [w], and all the vowels. All other sounds are referred to as *voiceless*. By placing the fingers lightly on the larynx, tactile feedback can verify whether the "voice box is turned on," a phrase often used to cue the voicing feature.

What are consonant cognates?

The feature of voicing distinguishes between certain paired phonemes that are referred to as *cognates*. These sets of phonemes are comprised of two sounds that are produced exactly alike, except that the vocal folds vibrate for one and not the other. The phoneme pairs are [p,b], [t,d], [k,g], [s,z], [ʃ,ʒ], [f,v], [θ,ð], and [tʃ,dʒ], with the first phoneme in each pair being the voiceless sound and the second the voiced phoneme. A fair amount of time should be spent helping children learn the quiet versus noisy twin because the cognates represent a high percentage of phonemes in the language, over one third, and the twin concept provides a cue for coding.

What can confuse children spelling cognates?

In other languages, phonemes can be distinguished based on partial voicing. Although English does not have separate phonemes for partially voiced or partially unvoiced phonemes, in fact, voicing in ongoing American English speech is not always an absent-versus-present feature. Phonemes are often altered for voicing by the phonemes surrounding them, so children spelling the wrong cognate may be perceiving a partial voicing feature and should be encouraged to listen more carefully to be sure that they are making the right discrimination decision.

Dental (Teeth)

How do [f] for [θ] and [θ] for [s] articulation substitutions sometimes affect spelling?

American English has four sounds that are produced with the teeth, [θ], [ð], [f], and [v]. Children who say *fink* for *think* and *think* for *sink* may have particular difficulty with the triad of sounds [f], [θ], [s] and their cognates. The phoneme [f] may be produced and perceived adequately, but since [f] is substituted for [θ] and [θ] is substituted for [s], the reference point for letter coding is a moving target for these children. They may not spell *thoup* for /soup/, but accurate coding of these sounds may take longer, they may make other letter substitutions, or they may just delete representation of the sound. Alerting them to the fact that the tongue is between the teeth for one set of dental sounds, [θ] and [ð], and the upper teeth touch the bottom lip for the other set, [f] and [v], but that [s] and [z] do not involve the teeth or lips can clarify the reference point for letter representation as they think about the sounds in the word they are trying to spell.

Plosive (Exploding, stopping, popping)

What are the two stages of plosive phonemes?

Phonemes with this distinctive feature have a popping quality produced by air buildup and sudden release. The stop (plosive) feature has two stages, implosive (buildup) and explosive (release). Depending on where the sound falls in ongoing speech, one or both stages of the explosion are used. Various articulators that participate in restricting airflow include the lips, tongue tip, alveolar ridge (bumpy ridge behind the gum that holds the upper front teeth), tongue back, soft palate, and pharyngeal (throat) wall. In American English, the stop sounds are [p], [b], [t], [d], [k], [g], [ð], [dʒ], and [ʔ] (glottal stop).

Where is the air restricted for [ʔ]?

The glottal stop [ʔ] is not always coded in SAE. Like [h], it is more like an approach to making a vowel sound and is sometimes heard between two vowels, for instance, in the word *kitten* or in words beginning with vowels, for example, *onion, utmost,* and *uh oh.* Children are sometimes confused about the pronunciation of [ʔ] in words and do not expect /t/ in spelling and reading notation, for example, *written*. The easiest way to present this letter representation is to say, "You are right; it does sound like another sound in the middle, not [t], when we say the word *written,* but this is one of those times when we have to pronounce the word for spelling instead of how we actually say it when we are talking." Demonstrate the artificial pronunciation, [wrɪtən], and the natural pronunciation, [wrɪʔn].

What two phonemes are paired in /qu/?

The phoneme patterns [ks] (*explain*) and [gz] (*example*) that are spelled by /x/ (and a few other patterns in higher grade lists) and [kw] that is spelled by /qu/ (and a few other patterns in higher grade lists) are treated in the plosive category, even though [ks], [gz], and [kw] represent two phonemes, a plosive plus sibilant for [ks] and [gz], and a stop plus glide for [kw]. Children typically tolerate this exception to NAR's requirement of coding one sound at a time in the block spelling sheets because /x/ and /qu/ words tend to be infrequent until the upper-grade spelling lists. At that point, children usually have enough meta thinking to find interest in or receive benefit from this peculiar set of spelling rules. Because the patterns are relatively unique, sometimes they are easier to recall once the pattern is unraveled.

What makes plosives easier to spell?

One beneficial feature of spelling plosives is that all but [k] have only two or three spelling patterns throughout all grade levels. All have single as well as double consonant representation of the phoneme, for example, /pp/ and /gg/. This is a cue that children should begin to consider when deciding which letter spelling would be correct, and it will become a recall hook for sorting options. Because fewer spelling options need to be considered for plosives, they often are a comfortable starting point in the NAR program. Double consonant spelling rules are probably too complex and have too many exceptions to present didactically as part of *NAR*. They are often based on syllable prosody but mostly on letter patterns. If a child is ready for this level of differentiation, several references are available on the Internet, and a few more frequent ones are presented in Appendix H. Language arts references tend to be more selective in the patterns presented, so cross-referencing sources to select what a particular child needs is important. A sample of possible points of clarification for doubling consonants can be found in Appendix H.

Sibilant (Hissing, snake, sharp, strident)

What causes sibilancy?

Sibilant sounds have a hissing quality that reflects mid- to high-frequency energy focus as the airstream passes over the forward-positioned and further back, grooved tongue and is directed toward the upper teeth. The child can be sensitized to the narrow opening between the tongue tip and roof of the mouth behind the upper teeth that produces the hissing quality for these phonemes. Contrasting a sibilant phoneme with one such as [ɑ] clarifies the air restriction. Sibilants include [s], [z], [ʃ], [ʒ], [tʃ], and [dʒ]. The affricate phonemes [tʃ] and [dʒ] are unique because they have both a plosive and a sibilant quality. At early NAR spelling-list levels, the spelling options for these phonemes are minimal. As higher grade-level lists become longer and have more complex words, spelling options are challenging. For instance, the syllable [ʃʌn] has numerous spellings at higher levels, but the suffix [ʃʌn] occurs so frequently that older children appreciate having explicit knowledge about the four or five spelling options regularly used in addition to /-tion/.

Fricative (Scraping, sanding, buzzing)

What is the difference between sibilancy and frication?

The fricative sounds are also high-frequency phonemes produced by constricting the airflow. The teeth and bottom lip restrict airflow for [f] and [v], the upper and bottom teeth restrict airflow for [θ] and [ð], and the vocal folds restrict airflow for [h]. Children who present a fricative simplification phonologic process when writing *teef* for /teeth/ or *free* for /three/ need clarification of this feature. Sometimes they will say *free* for *three* as they are spelling, but spell /three/ correctly, another sign of emerging discrimination. The glottal fricative [h] does not really function as a phoneme in many instances but is sometimes heard in transition to and from phonemes. The glottal fricative is present at the end of a vowel sound as the vocal folds stop vibrating and air begins to flow over the vocal folds again, but it is generally not perceived or designated as a phoneme. Some children, however, may spell /h/ at the end of a syllable ending in a vowel sound because they actually hear the [h], so they should be credited with good listening. In other instances, the glottal fricative is registered and coded as a phoneme, for instance, [h] in *hand*. The fricative feature of [h] is redundant for classification because the other fricative phonemes are distinguished from it by dental involvement and other unique features of their bundles, but training for perception of the frication feature can be useful because it provides another cue for children to sort auditory information.

Nasal (Nose, honking)

What closes the oral passageway for nasal phonemes?

Only three phonemes are resonated in the nasal chamber in SAE, [m], [n], and [ŋ]. All other sounds are resonated in the oral chamber. Even though these phonologic rules generally apply, many children and adults produce other sounds, especially certain vowels, with varying degrees of nasal resonance. Mixed oral and nasal release of vowels can sometimes yield misperceived words, but usually speech is readily understood and just referred to as nasal. However, a young child learning to spell with an adult exhibiting hypernasality could have more difficulty discriminating. To help clarify the airflow patterns, lightly touching the nose will allow the child to feel the resonance in the nasal chamber, or placing a mirror under the nares will reveal steam on the mirror for nasal phonemes. To produce the nasal feature, the voiced airstream is not sent through the oral cavity as it is for all other SAE phonemes. Instead, the velopharyngeal port remains open, redirecting the air upward behind the blockage. Conditions such as nasal congestion, which blocks airflow through the nasal chamber, has a significant impact on nasal versus oral sounds. A cold tends to make nasal phonemes sound oral. If blockage is severe and pressure to release a sound is maintained, plosive phonemes [p], [t], and [g] can be heard in place of [m], [n], and [ŋ], respectively. If the air is allowed to escape into the nasal chamber, all phonemes have nasal resonance, which creates the quality heard in individuals with cleft palate.

What are the variables that determine nasal resonance?

The force of airflow, effectiveness of the valves that direct air flow at the velopharyngeal port, and the anatomical structure of the laryngeal, oral, and nasal chambers determine the degree of pressure necessary to accomplish nasal or oral resonance. Individuals present their own set of dynamics, partially accounting for the fact that the speaker can be identified by voice quality, even when not seen. Although hypernasal resonance is generally avoided if it can be, it occurs unintentionally and intentionally in certain singers and genres of music—with mixed reviews.

What nasal consonant pattern yields frequent misspelling?

The fact that nasal leakage is often heard in normal speech may have something to do with the difficulty many children have in distinguishing nasal sounds for letter coding or the difficulty perceiving nasality could be due to other acoustic features, but it often occurs. Nasal sounds are often deleted in spelling, particularly when they are preceded or followed by a phoneme that has a similar mouth position, so that *stamp* may be spelled as *stap*, *spend* as *sped*, and *went* as *wet*. It is important that children become sensitized to the nasal feature to clarify spelling.

Glide (Sliding, moving, joining)

How do glides differ from consonants and vowels?

Unlike other consonant and vowel sounds, glides are produced *while* the tongue, lips, and jaw are in transition from one phoneme to another. The four glides of American English are [w], [r], [j], and [l]. Three of the glides begin or end at a specific vowel area. The glide [j] begins or ends in the [i] vowel area, the glide [r] begins or ends in the [ɝ] vowel area, and the glide [w] begins or ends in the [u] vowel area. Children with dyslexia and adults with auditory discrimination deficits often perceptually merge the glides with surrounding phonemes. The confusion is typically detected in deletion and sometimes in addition, substitution, or transposition of letters representing these phonemes in spelling and reading. Asking children to feel the tongue move from [i] to [u], producing [j], as they hear the word *you* can demonstrate the movement. Then they can move from [u] to [i], producing [w], to hear *we*. Exercises based on words showing these patterns can help clarify the movement nature of glides and facilitate accurate letter coding. Glides are neither consonants nor vowels and are sometimes referred to as *semivowels* because of their close association to vowel positions.

What is unique for discriminating [l]?

Some consideration should be given to the glide [l] in unstressed middle syllable and final word position. Spectrographic analysis of words such as *pencil*, *satellite*, *table*, *racial*, and *principal* show slight variations in how the [l] element is pronounced, but the differences are subtle and too difficult for most to decipher. Sometimes the [l] ending seems to be preceded by an [ə]-like sound, other times by an [ʊ]-like sound, and other times as a "dark" [l] produced farther back in the mouth. NAR approaches the dilemma by grouping all these variations as one phoneme with seven options for spelling [l] by fifth grade.

Where does air emission take place for [l]?

The glide [l] is unique because it is the only lateral sound in the language, meaning that air escapes over the sides of the tongue rather than out the front. Some people have a lateral lisp in which this feature is erroneously produced for sibilant sounds, giving speech a slushy quality.

What is the role of letter vowels in coding [r] and [ɝ]?

Once the child perceives the transient feature of [r] and the longer duration of [ɝ], some rules of letter coding can be introduced. A useful coding strategy is to consider that the vowel [ɝ] is never spelled /r/; it always has one or more vowel letters that "help" it be [ɝ], for instance, /ir/, /er/, /ear/, /irr/, and several others. Letter vowels surrounding [r], by contrast, make their own sound, for instance, /car/, /carrot/, and /around/. This distinction helps clarify many spelling and reading errors. In traditional reading activities, many who model production of the two phonemes, including computer programs, do not distinguish between the held-in-place phonation feature of the vowel [ɝ] and the transition feature of [r]. In fact, they prolong saying [r] so that it actually becomes [ɝ]. This especially confuses children when they are already struggling to understand phonics terminology such as /r/ "syllable patterns." The instruction does not correspond to the physics and acoustic parameters necessary for the distinction between [r] and [ɝ]. It cannot be emphasized enough how frequently the [r] and [ɝ] phonemes occur in SAE and how difficult their management can be.

Which is easier to spell, [r] or [ɝ]?

The reason that children need to be taught the distinction between [r] and [ɝ] is that the spelling differences are extreme. The phoneme [r] has 4 spelling options in the fifth-grade list, and the phoneme [ɝ] has 10, plus the rules for vowels on either side of the [r] are much different than for the vowels accompanying [ɝ].

Why is [ɔr] treated differently?

When the phoneme preceding [r] is the short-duration vowel [ɔ], children often hear the combination as one "sound," so NAR includes a list for that combination, making sure that they hear both the [ɔ] and the [r] elements.

What is the difference between [ɚ] and [ɝ]?

Because the [ɚ] versus [ɝ] distinction involves stress detection and is often difficult for children to identify, words with the stressed and unstressed forms of [ɚ] and [ɝ] are merged into one list in NAR, but [r] lists are necessarily separate.

What is unique about [ɑr] words?

Children often incorrectly spell words with [ɑr] phonemes in them, such as *car, far, bar, archer, bark,* and *park,* spelling them as /cr/, /fr/, /br/, /rcher/, /brk/, and /prk/, respectively, because the letter name of /r/ is [ɑr]. This reflects visual letter name dependency in coding, a signal that should be monitored.

What placement differences can be detected?

Charts showing mouth placements for different phonemes are often helpful in demonstrating place features of articulation, such as those found in *Applied Phonics: The Sounds of American English* (Edwards, 2002). By comparing charts of phonemes with subtle differences, for instance, between [r] and [ɝ], children can sometimes crystallize their understanding of phoneme differences. Most clinicians will have several sources for replicated or stylized phoneme mouth postures. A few diagrams in Appendix I may be helpful.

What are allophonic variations?

Glides represent a unique timing feature of articulator movement during phonation that defines the phoneme, but all phonemes are affected by movement from one phoneme transition to the next in words, referred to as the coarticulation effect. A phoneme is actually comprised of many allophonic variations that are determined by the adjacent phonemes. These variations come to be recognized as one phoneme. The influence of surrounding sounds can even extend beyond the phonemes immediately preceding and following a phoneme. This can be noted when saying the word *blue*, for example; the lips begin to round for [u] when the [b] is produced, even before [b] or [l] has been said. Comparing the [b] movements for *blue* to those for *ball* clarifies the early start for *blue*. These preparatory postures that are not synchronized with the actual production of a future phoneme can confuse perception but can be resolved by isolating phoneme elements in the word. Preparatory mouth postures and transitions to and from previous and next phonemes are explicitly presented in "Dynamic Aspects of Speech Production," in *Phonetics: Principles and Practices* (Singh & Singh, 1982, Chap. 6, pp. 83–171). The authors used filmstrips depicting coarticulation of phonemes in several words, with transitions reflected in a series of pictured mouth changes over time. These pictures display the fluidity and complexity of phoneme production that children learn to ferret out and perceive as individual phonemes and then code with letters as words.

Vowel Phoneme Distinctive Features

What shapes vowel differences?

As discussed previously, vowels have less airflow restriction from articulators than do consonants. Vowel sounds are still shaped, however, by slight restrictions produced by lip rounding or spreading, tongue height, tongue forward or back placement, and jaw position. Using other senses, such as vision and touch, to teach these more subtle characteristics to children who cannot auditorily discern differences between vowels can be helpful. Sometimes just discussion is sufficient, but more often children appreciate seeing or feeling the features of vowels that distinguish each one from others.

How does English manage to spell 20 vowel phonemes with only five vowel letters?

SAE has approximately 14 single-phoneme vowels and six diphonemes (two-phoneme vowels), with only five letters to code all of them. Children must learn an astounding number of letter combinations with duplicated and overlapping patterns to represent these different vowel patterns. This represents yet another layer of rule coding that young spellers and readers must conquer.

What is problematic about long and short vowel designation?

Vowel duration is the primary classification of vowels in traditional reading instruction. Of the approximately 20 SAE vowels, educators designate 5 vowels as long; for instance, [eɪ] is called long /a/, [i] is called long

/e/, [aɪ] is called long /i/, [oʊ] is called long /o/, and [ju] is called long /u/. Four of the five long vowels taught in reading instruction are actually diphonemes, or two joined vowels. Other single-vowel phonemes are also long; for instance, [ɝ] and [ɑ] (or [a] in New England dialect) are long, but they are not classified as such in traditional reading instruction. Vowel length varies significantly both within different words and in comparison to each other as can be discerned in Table 7–2.

Considering that vowels are produced four to five times faster than it takes to blink and that the difference between a long and short vowel duration may be negligible, phoneme durations represent such miniscule time lapses that auditory perception of relative length is difficult. Vowel length is affected by variable contextual factors such as syllable stress and the type of adjacent phoneme. Adults may be able to perceive some temporal feature of length that the binary designation of long and short is intended to illuminate, but the duration classification is still in question when length is designated for only 10 of the 20 vowels of English. Vowel length described as *long* and *short* would probably best be deleted; however, it is so engrained in traditional instruction, including the dictionary pronunciation key, that it must be taught around by providing children supplementary information. Children often show confusion about vowel length; for instance, they may hear two vowels in [aɪ] and want to add a letter(s) to spell the diphoneme, spelling *ice* as /aees/. Again, they should be praised for their good listening, but explaining the length issue would probably be counterproductive.

What does the term *diphthong* mean in terms of visual versus auditory coding?

Just the concept of diphthong can confuse children. Some use *diphthong* to refer to two joined vowel letters, while others refer to *diphthong* as two joined vowel phonemes. In spelling, the number of letters that spell phonemes can vary considerably; sometimes two letters make one phoneme

Table 7–2. Phoneme Durations

Phoneme	Duration (in msec)	Phoneme	Duration (in msec)
[i] eat	78–119	[ɑ] father	93–140
[ɪ] it	8–75	[ɔ] or	72–148
[e] vacation	78–136	[o] hobo	96–162
[ɛ] every	60–106	[ʊ] book	61–85
[æ] at	71–159	[u] shoe	75–126
[ə] above	49	[aɪ] eye	114–172
[ʌ] above	103	[aʊ] out	202
[ɝ] further	120	[ɔɪ] boy	290
[ɚ] further	96	[ju] unit	270

Source: Compiled from Edwards, H. T. (2002). *Applied phonetics: The sounds of American English* (3rd ed.). Clifton Park, NY: Delmar, Cengage Learning.

as in /ee/ for *see* [si]; sometimes one vowel letter makes two phonemes, for example, /i/ in *hi* [haɪ]; sometimes two letters make two phonemes as in /ow/ in *know* [noʊ]; sometimes three letters make one phoneme, for instance, /ear/ in *early* [ɝli]; and sometimes four letters make one phoneme, for example, /ough/ in *thought* [θaʊt]. The list goes on. Diphthong letter classifications are taught to children as though they predict pronunciation; for instance, "When two vowels go walking, the first one does the talking," which presumes each vowel letter has its own pronunciation, which it obviously does not.

What is a diphoneme?

Because the letter rules do not really match pronunciation, the only choice children have is to memorize the letter terminology and definitions and not dwell on the explanations. NAR uses the term *diphoneme* for two joined vowel phonemes because the term, at least, clarifies the designation as two phonemes, but the quantification of letter versus phoneme elements remains problematic.

What should children memorize for spelling?

Teachers note with frustration that children have so much trouble with the vowels. Of course they do! Children with dyslexia, who often do not perceive vowels accurately, especially fall prey to confusion with this type of rule teaching, or they learn the rules only to find that they cannot depend on them and become frustrated. The time and energy spent learning inaccurate or unreliable rules would be better spent accepting, for example, that the phoneme [i] (as in <u>ea</u>t) has nine ways of being spelled at a third-grade level lexicon. Just learning the spelling patterns of heard words by grouping them, talking about them, detecting patterns, using visualization and mnemonics for recall, and applying dozens of other association strategies would actually build spelling knowledge. Accurate phoneme perception is the first and most important step in spelling (and reading). Any coding system that does not begin by stabilizing accurate perception of phonemes and, instead, presents letter patterns to pronounce words is simply placing an unreliable, external set of rules on the child, not building on what the child already knows.

What does the vowel diagram represent?

NAR demonstrates that vowels can be learned based on how they sound and how they are made. The vowel diagram in Figure 7–1 references physical mouth features in relation to tongue positions for various vowel phonemes to guide the learning process. Vowels on the left represent the front of the mouth and, on the right, the back of the mouth. The vowels at the top of the diagram are phonemes that are made with the tongue, at its highest point relative to other vowels, partially restricting airflow. Vowels at the bottom are phonemes made with the tongue at its lowest point of partial restriction, in other words, leaving a larger space between the tongue and the top of the mouth. Other features shape the acoustic qualities of vowels, including jaw position, lip rounding, lip spreading, and mouth opening, and they are correlated with this top-bottom, front-back delineation of physical tongue placement.

These and other features are described in more detail. Keeping the vowel diagram in sight while considering structure and movement features

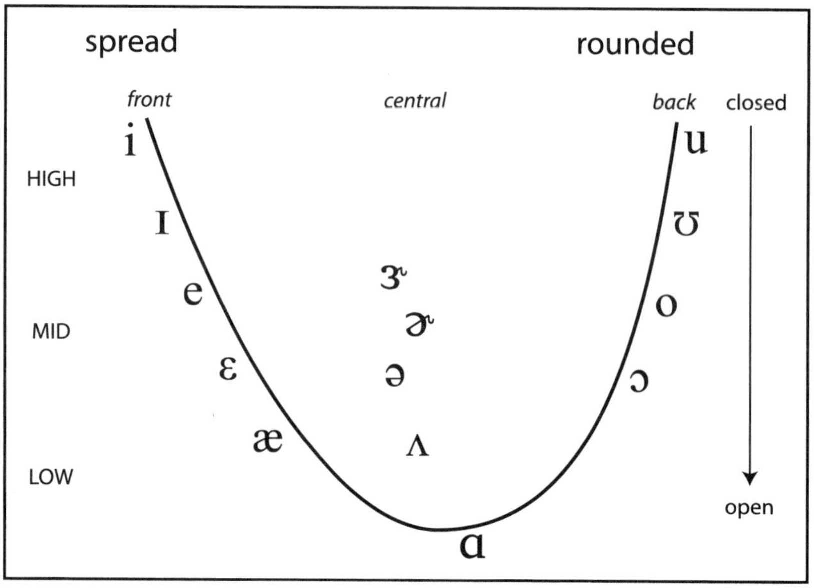

Figure 7–1. Vowel diagram.

clarifies the patterns. Children become proficient in using this diagram or simplified versions of it, with reference to their own mouths, and often visualize the schematic to help decipher ambiguous phonemes. Examples of activities to clarify features can be found in Appendix C. It is impossible to avoid the reality that spelling vowels is difficult, but it is possible for children to perceive vowel differences using several scaffolds representing the physics, acoustics, and anatomic features of phoneme production to help the discrimination, recognition, and identification processes. Hearing and feeling these fine differences is a first step to prepare for the dreadful task of letter coding.

Lip Rounding Versus Spreading

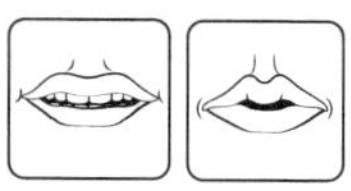

What are the polar lip postures for vowels?

Front vowels such as [i] and [ɪ] are produced with the lips spread, with the spreading gradually increasing as the tongue is raised. Others, such as [u], [ʊ], and [o], are produced with the lips rounded, with the rounding gradually increasing as the tongue elevates for the back sounds. The sounds [e], [ɛ], [æ], [ɑ], [ə], and [ʌ] are produced with the lips in a neutral position. This is an easy-to-see distinctive feature of vowels, especially at the extremes of [i] and [u]. Children can look in a mirror and readily see the differences.

High Versus Low Tongue Height

What part of the tongue peaks for vowels?

When terms such as *high, middle*, and *low* are used to describe tongue positions for various vowels, they refer to the highest point of the slightly arched tongue, not the tongue tip. This can be a difficult distinction for children to understand because the body zone of the tongue (behind the tip) is difficult to see, and that is where the tongue lifts. Much of the instruction about this feature of the tongue needs to emphasize how the tongue feels in relation to diagrams that map the highest part of the tongue for each vowel. Some visual information can be obtained from looking in a mirror or in someone else's mouth using good illumination. The vowels [i], [ɪ], [u], and [ʊ] are considered high; [a], [ɑ], and [æ] low; and [e], [ɛ], [ɔ], [o], [ʌ], [ə], [ɚ], and [ɝ] middle.

Open and Closed Vowels

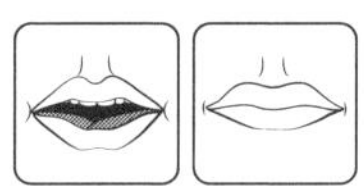

What aid to vowel discrimination does the jaw provide?

The jaw carries the tongue to high, low, and middle vowel positions. Since the jaw can be seen much more easily than the highest part of the tongue that determines the vowel being produced, the jaw can be used as an indicator of tongue position. The jaw is closed for the vowels [i], [ɪ], [u], and [ʊ]; open for [æ] and [a]; and half-open for [e], [ɛ], [o], [ɔ], [ʌ], [ə], [ɚ], and [ɝ].

Front Versus Back Vowels

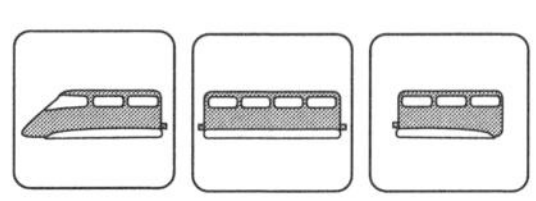

What are the farthest front and back vowels?

The highest part of the tongue is in the forward position to produce vowels such as [i], [ɪ], [e], and [ɛ]. Other vowels, such as [u], [ʊ], [o], and [ɔ], are made with the highest part of the tongue toward the back of the mouth. Central vowels are those made with the highest part of the tongue in the middle of the mouth and include [ɑ], the schwa sounds [ə] and [ʌ], and the stressed and unstressed "er" sounds [ɚ] and [ɝ].

What effect does syllable stress have on vowel integrity?

The schwa represents the vowel sound that is made when the mouth is in rest position, and it can occur in a stressed or unstressed syllable. Syllable stress for word pronunciation is another of the more difficult perceptual tasks for many spellers with dyslexia to master. SAE stress patterns for polysyllabic words do not have dependable rules that designate which syllable should be stressed in two-syllable or polysyllabic words; however, one vowel pattern in relation to syllable stress is consistent: As a syllable receives secondary or tertiary stress, the vowel in those syllables tends to gravitate to the neutral (or schwa) position.

References

Blache, S. E. (1978). *The acquisition of distinctive features*. Baltimore, MD: University Park Press.

Edwards, H. T. (2002). *Applied phonetics: The sounds of American English* (3rd ed.). Clifton Park, NY: Delmar, Cengage Learning.

Hall, D. C., & Mielke, J. (2014). *Distinctive features.* Retrieved from http://www.oxfordbibliographies.com/view/document/obo-9780199772810/obo-9780199772810-0012.xml?rskey=kMoJ8q&result=32

Hertz, S. R., Younes, R. J., & Zinovieva, N. (1999). *Proceedings from 14th International Congress of Phonetic Sciences: Language-universal and language-specific components in the multi-language ETI-Eloquence text-to-speech system.* Retrieved from http://www.novaspeech.com/documents/ICPhS99.pdf

Miglietta, S. (2010). *Why we need distinctive features? The role of distinctive features I natural languages.* Retrieved from http://www.cril.unile.it/download/SandraMiglietta_WHY_WE_NEED_DISTINCTIVE_FEATURES__09-07-2010.pdf

Singh, S., & Singh, K. S. (1982). *Phonetics, principles and practices* (2nd ed.). Austin, TX: Pro-Ed.

Tatham, M. (1999). *Distinctive feature theory.* Retrieved from https://www.msu.edu/course/asc/232/DF/df-theory.html

White, L. (2003). *Second language acquisition and universal grammar*. New York, NY: Cambridge University Press. Retrieved from http://bilder.buecher.de/zusatz/23/23181/23181107_lese_1.pdf

8

Detailed Description of Stage III: Coding Phoneme Patterns With Colored Objects

Color Coding Phoneme Strings

What advantages does a nonword color-coding activity have?

The word lists in Appendix J used in Stage IV, the letter-coding stage of Neurolinguistic Approach to Reading (NAR), can be used for color coding, or word lists can be gathered from other print sources used by the child. Pictures of words can solidify the reference being coded. Many commercial products are available for this stage of color coding, such as Phonological Awareness Training for Reading (Torgesen & Bryant, 1994) and The Phonological Awareness Kit (Robertson & Salter, 1997). For older children who still need this type of coding on more complex words, and most do, pictures of nouns and verbs can be presented without saying the word, which gives opportunity to observe how well inner language coding is progressing. Several products are available that provide nonword lists targeting various syllable patterns (Lindamood & Lindamood, 1998; Rosner, 1986; Sloan, 1991). A few nonword lists are found in Appendix D, Tables D–1 through D–3, that are tailored to address phonologic processes that tend to give particular difficulty to individuals with dyslexia. These nonwords can be coded as first presented and then as altered for phoneme order, omission, addition, and transposition, drawing on increasingly complex listening skills. Clinicians can design their own nonword lists targeting certain features, phonemes, or classes of phonemes that the child is having trouble perceiving. These lists can be particularly useful for children who have been sufficiently exposed to reading that they have visually memorized word spellings and yet do not accurately perceive the phoneme elements in sound strings. Other exercises are provided in Appendix D, Tables D–4 through D–9, as examples of auditory training work.

Table D–1 is a phoneme deletion exercise. Table D–2 is a two- and three-syllable exercise for identification of a numbered phoneme, for example, third phoneme. Table D–3 allows discrimination of interdental, labiodental, and sibilant phonemes. Table D–4 provides words for pictured distinctions of [ɪ], [ɛ], and [æ], as does Table D–5 using nonwords. Table D–6 allows charting of progress for exercises. Table D–7 addresses front vowels. Table D–8 provides glide contrast words, and D–9 challenges nasal discrimination.

The rationale for increasing children's use of phoneme analysis when they have already sight-read a word correctly is that they are not taking advantage of the auditory features of words that trigger vocabulary recognition, which, in turn, facilitate the semantic association process for contextual meaning. Furthermore, they are not using strategies that will allow them to decipher unfamiliar words.

How are colors used to code words? How do color coding errors cue a confused feature?

NAR lists in the grade-level spelling sets are based on words rather than nonwords. Words are spelled using colored objects, for instance, flat marbles, blocks, miniature animals or shapes, play gemstones, tiles, or any set of small objects that have at least five duplicates of six or more colors that can designate the number and sequence of phonemes in the string, whether phonemes in the string are the same or different, and whether any phonemes are duplicated. At higher levels, more colors may be needed if a child is tackling particularly long, polysyllabic words. Coding is usually not color specific, that is, [t] is not always white and [æ] red, but, for children with adequate short-term memory, short phoneme strings could be coded using a color key. Examples of coded strings for [d] words using different-colored chips in the third-grade list are displayed in Table 8–1.

Table 8–1. Color Coding Phoneme Strings

Word	First Sound	Second Sound	Third Sound	Fourth Sound	Fifth Sound	Sixth Sound
muddy	[m]	[ʌ]	[d]	[i]		
	/m/	/u/	/dd/	/y/		
	red	green	black	white		
monkey	[m]	[ʌ]	[ŋ]	[k]	[i]	
	/m/	/o/	/n/	/k/	/ey/	
	white	green	blue	yellow	red	
indeed	[ɪ]	[n]	[d]	[i]	[d]	
	/i/	/n/	/d/	/ee/	/d/	
	brown	red	blue	green	blue	

How can blending phonemes and segmenting words be used with color coding?

This stage of coding should be continued until the words are coded quickly and accurately and no particular phoneme, class of phonemes, or phoneme feature interferes with the fluency of object coding. Any patterns of perceptual confusion should be addressed by returning to Stage II activities that target the vulnerable phoneme characteristic. For instance, if [t] and [d] are coded with the same color, Stage II activities that address *voicing* should be reviewed; if [n] is not coded, *nasal* clarification is needed; if [s] and [ʃ] sounds are coded with the same color, *rounded* and *front* feature discussion and demonstration activities are needed; and if [f] and [θ] are coded with the same color, *frication*, *labial*, and *dental* clarification is needed. In other words, the errors signal what features of the phonemes under consideration are confused and what clarification is needed, allowing for the fact that some errors are just random and do not relate to feature confusion. The clinician is always looking for patterns.

References

Lindamood, P., & Lindamood, P. (1998). *The Lindamood phoneme sequencing program for reading, spelling, and speech* (3rd ed.). Austin, TX: Pro-Ed.

Robertson, C., & Salter, W. (1997). *The phonological awareness book—Intermediate*. East Moline, IL: LinguiSystems.

Rosner, J. (1986). *The green readiness book: Auditory and general readiness activities for reading and arithmetic*. Austin, TX: Pro-Ed.

Sloan, C. (1991). *Treating auditory processing difficulties in children*. San Diego, CA: Singular.

Torgesen, J. K., & Bryant, B. R. (1994). *Phonological awareness training for reading*. Austin, TX: Pro-Ed.

9

Detailed Description of Stage IV: Spelling Sounds With Letters

How is alphabet spelling an invented system?

As critical as it has been to build strong phonologic perception skills to succeed in written language acquisition, children must learn more. They must learn an English spelling system that is one of the most complex systems of any language. Until such time that organizations such as the Simplified Spelling Society (http://www.spellingsociety.com) succeed in their mission to change English spelling, which is predicted to be no time soon, if ever, children must grapple with the intense memory task of English spelling irregularity. Many reading programs try to make children think that they must learn rules about how letters should be pronounced, but learning those rules is a great memory burden. Furthermore, the rules are really patterns with so many exceptions that they confuse the process for many children, especially children with dyslexia. Children must face the fact that phonemes have many spelling options. The patterns of phoneme spellings, not letter pronunciation patterns, need to be presented systematically. This allows written language acquisition to begin naturally, with phonemes, as language was learned in the first place, not as the invented system of spelling that has been superimposed on that developmental process and does not actually fit. Spelling irregularity only compounds the already weak phonologic perception that has blocked reading acquisition early on for children with dyslexia.

Should children practice hearing or seeing the word before they spell it?

Even as children begin to consider the correct invented visual symbols involved in spelling and reading, they must use phonologic perception to guide the process and to make sure that the visual symbol read or written is a correct fit. At this stage of Neurolinguistic Approach to Reading (NAR), for the first time, visual symbols, letters, are the reference for coding. Visual recall of correct letter spelling patterns is encouraged for match to heard phonemes. Once auditory perception for Stage II activities is fast and accurate and Stage III phoneme color coding is fast and accurate, children are ready to apply letter coding to what they already know. The critical principle to be employed at this stage is that children always hear the word before they are asked to spell it.

Why can spelling skills plummet by third grade?

NAR spelling strategy is unlike school spelling procedures in which the child takes home a list of printed words on Monday, memorizes them, and takes a test on Friday. Children using the NAR program do not see the letters of the words they are to spell until they hear them, repeat them, and spell the word, phoneme by phoneme, with letters. They write letters based on the phoneme patterns they hear. The objective is to increase auditory memory, not visual memory of word patterns until auditory perception is strong enough to guide recognition of previously known words as well as unknown words. In other words, this program builds spelling proficiency before it expects reading proficiency. Spelling must be supported with phonologic perception or it fails when the vocabulary becomes too large to memorize visually. As Moats (2005–2006) points out, after third grade, spelling skills can significantly drop in vulnerable children compared to comprehension skills. This is around the point at which vocabulary is expanding so rapidly that dependency on sight reading begins to exceed ability to visually recall word letter patterns. This is a critical stage for many children and often is the point at which dyslexia is diagnosed, not because they suddenly came down with a reading disorder, but because now the task has become impossible and unexpected frustration behaviors emerge in school and at home when reading. The behavior of children who had seemed smart and resilient can change after first exposure to reading instruction to become sullen, angry, or depressed. Parents are often confused about the change and are trying to find the source of the problem. Sometimes they hold the child accountable for behaviors, not understanding how embarrassed the child has been made to feel. As Lyon (1997) gathered data on dyslexia and reported in "Why Johnny Can't Decode," he wrote that the embarrassment and humiliation children with reading problems display increasingly saddened him and that children were not as resilient as he once had thought they were.

How were NAR spelling lists constructed?

If children do not need nonsense words to avoid sight-reading strategies, it is better for them to devote their cognitive energy to spelling actual words from the vocabulary they will face in school. The NAR lists in Appendix J were collated from approximately 10 grade-designated spelling resources using a merge-purge sorting process. The words were then reviewed for idiosyncratic words unique to a specific context and not universal to a grade level. Each word was then assigned to a phoneme category represented by one of the phonemes in the word. Consideration was given to balance in phoneme representation so that each phoneme list had a similar number of words or a sample representative of the occurrence of the targeted phoneme in the language.

How is an appropriate spelling list selected?

Selecting the appropriate grade-level set of lists for a child is an important step at this point in the program. If reported reading level were a viable concept for the child with dyslexia, beginning at grade score designations would be the logical first consideration. However, most grade scores are based on an average of coding and comprehension ability, with coding inordinately weak and comprehension significantly higher in most children with dyslexia, pulling the grade score up. Also, grade scores are statistically weak and can be misleading. To be at a comfortable task level, the list should target coding scores on standardized oral reading tests measuring the rate and accuracy of oral reading. Consideration should be given to the age and cognitive ability of the child so that vocabulary in the list is not too limited or advanced. If a very bright child learns the instructional process and the spelling options quickly, it may be important to switch to a higher grade-level list to maintain motivation. The reason that a great deal of elaboration on word meaning is usually not necessary is that typically coding skills are below receptive and expressive vocabulary levels, and as soon as the string of phonemes are recognized as a word, the meaning is available.

How quickly should a child master a grade-level list?

As mastery is achieved at lower levels, a higher grade-level set can be introduced, even before completing a lower grade-level set if the child is doing well. The child can skip phonemes that were done quickly and accurately in the previous level, but the clinician should check for any new spelling option that was not in those previous lists that will need to be practiced in the new grade-level set. Some children take a long time to master lists. One child may take 3 years to master one grade set while another may take 1 year to master three grade sets. This is typically related to the severity level of the dyslexia, but other factors are operational as well.

What purpose do high school lists serve if they are so difficult?

Exceptionally bright children with dyslexia often know and attempt to use vocabulary found in the higher grade-level sets, but their spelling levels are so far below their lexicon levels that their written expression does not match literacy expectations. Spelling lists beyond fifth grade are important because they provide these children a goal resource. Higher level vocabulary may better represent their cognitive ability, but the high school grade-level sets are challenging for most children and adults with dyslexia. Many will find them too challenging to master, but they should still see improvement.

What two goals does NAR seek to meet?

Although stopping NAR when grade-level scores are achieved on nationally standardized tests may seem logical, this may be too simplistic a criterion for discontinuing spelling and reading support. As previously noted, children with dyslexia present a perplexing picture of significant strengths and significant weaknesses. For children who are gifted and talented and whose reading comprehension may be improving to be more in line with their intelligence level, it is important to be certain that their coding skills are commensurate. If the coding lags, the child will probably continue to avoid reading and, when it is required, will show resistance. The goals of NAR are to decrease the discrepancy between coding and comprehension scores in timed, paragraph oral reading measures and better align reading level with cognitive ability. When these two outcomes are met, competent and comfortable spelling and reading should be reflected in written language tasks at home, school, work, and social events.

How are NAR lists organized?

In each grade-level set from kindergarten to 13th grade, the phonemes are introduced with reference to a class of phonemes, that is, *plosive, sibilant, fricative, nasal, glide, vowel*, and *diphoneme*, phoneme classifications that have been part of the learning process since Stage II. By Stage IV, children are ready to spell words with consideration of a target phoneme that falls in a given class. Each grade-level set follows the same sequence of classes and phonemes within classes, which helps in finding a particular phoneme or class of phonemes when moving from one grade set to another. Each list of words demonstrates the variety of ways the target phoneme is spelled at that grade level. This is an important feature for younger children, who can be overwhelmed by older grade-level lists that have many more options for spelling different phonemes. If they begin with a list of limited spelling options that matches their memory capability to learn, it is much easier to increase their spelling options gradually as they take on older lists. This is one of the most important reasons for early intervention; it allows a more manageable number of early words to learn that can be gradually increased. This ideal scenario of early intervention, however, is not typical, and children still manage to grasp spelling options with NAR, albeit with more difficulty. And the difficulty of the task needs to be taken seriously. Improving spelling and reading will probably require more repetition and practice than the child would like, but it should also be enjoyable and not too demanding. It is important that the child be on board, perhaps with a significant reward waiting for reaching a significant goal. What better accomplishment deserves reward than significant improvement in written language?

How can parents, siblings, tutors, and others help habituate spelling patterns?

The adult should decide what percentage of words needs to be spelled correctly to know that the child has mastered the spelling options on a list. Some children with weak phonologic memory may have to practice lists several times before they are comfortable recalling all the spelling options for a given sound and the exact spelling patterns for each word on the list. Other children will need to practice the list fewer times. Most children will require additional activities to solidify the spelling patterns associated with each phoneme (see Appendix J for examples). At this point particularly, parents, siblings, and tutors can help solidify storage of correct spelling patterns. Respelling previous errors can be charted to show progress in the percentage of correctly retained patterns. Seeing the gains can encourage a child.

How can threshold of sight reading be detected?

Children who sight read and have good visual recall of letter patterns for spelling but who still struggle in sounding out new words will probably be the children who seem to have no difficulty on earlier grade-level sets. As they move to more difficult sets containing unfamiliar words that preclude sight guessing or overrely on visual recall, the phonologic element in reading breakdown will surface. At that point, the clinician can address the breakdown by inserting information from previous stages in the program and introducing some of the pronunciation patterns that need clarification for spelling.

References

Lyon, G. R. (1997, October 27). Why Johnny can't decode. *The Washington Post*. Retrieved from http://www.ldonline.org/article/6395

Moats, L. C. (2005–2006). How spelling supports reading: And why it is more regular and predicable than you think. *American Educator*, pp. 12–43. Retrieved from http://www.aft.org/pdfs/americaneducator/winter0506/Moats.pdf

10

Special Considerations

Pronunciation Peculiarities

What governs pronunciation of /t/ as [d] in middle word position?

American English speakers usually pronounce words spelled with /t/ as [d] when in an unstressed syllable in the middle of words so that *started* is pronounced *starded, thermometer* is pronounced *thermomeder*, and *battery* is pronounced *badery*. These unique features of pronunciation make auditory discrimination for certain words more difficult, and adults need to guide children's spelling. Again, just pointing out the peculiarities empowers children to unravel spelling. The website http://www.americanaccent.com/pronunciation.html provides additional explanation for pronunciation patterns for the letter /t/ in Standard American English (SAE). Perceiving and producing this distinctive feature rule of SAE tends to be one of the last pronunciation patterns English language learners (ELLs) master. Their accent maintains pronunciation of [d] as [t] because they are paying more attention to the letter than the way Americans pronounce these words. Children with reading and spelling problems are also often confused by this rule as they spell *pardy* for /party/ but *tarty* for /tardy/! The strategy that seems to work best is, once again, to refer to the way certain words are pronounced for spelling compared to how they are said when talking. They do not need to know the syllable stress rule because it is at a rule level of language they do not need to know, and it has too many exceptions, but they do need to understand that something caused their confusion and that it can be bypassed.

Dealing With the Schwa

Why is the schwa particularly difficult to spell?

The [ə] is not just an unstressed form of [ʌ]; it carries some of the coloring of the vowel it would be in a stressed syllable. So the unstressed schwa is really a combination of many sounds that hover around the neutral mouth position but still has qualities of the source vowels. Eventually, the schwa will be spelled with many vowel letters or letter combinations. This means that the child must recall correct vowel spelling from long-term stored visual memory. To help do this, the child is taught to use a deduction process involving root word, prefix, and suffix patterns and to consider other words with similar sound patterns. These strategies are gradually introduced as words with these features emerge in progressing grade lists. Spelling the vowels in unstressed syllables is one of the most difficult spelling tasks.

What strategies can be used to help spell the schwa phoneme?

A compensatory strategy to help children perceive the phoneme they are coding is to pronounce the word as though the schwa were the source vowel and then store this pronunciation in memory as a cue for spelling by repeating the artificial pronunciation many times on different exposures to the word. This strategy links auditory and visual storage and aids recall. To help remember a spelling pattern, many people automatically use this strategy; they pronounce the word the way it is spelled to help remember a spelling pattern, but children typically have to be taught this strategy to clarify their perceptual confusions. Using this inaccurate auditory reference for pronunciation of phonemes in words can add humor. Words often sound bizarre, which is part of the intention, to make the absurdity trigger recall. The root word strategy used to help unravel spelling of unstressed vowels is to identify the root word, analyze the spelling of the vowel in that context, and spell the unstressed vowel with that letter. This strategy will usually, but not always, yield the correct letter representation.

How does visualizing a word help spelling?

Some people insist that they must see a word to know if it is spelled correctly, apparently having had previous success in retrieving letter patterns from visual long-term storage with this strategy. Some clinical attention is given to this strategy of visual confirmation of spelling, even for children who have weaker visual memory because the spelling task, at this point, must be paired with auditory symbol recognition, and visualization is a traditional therapy tool used to strengthen recall. Expanding that visualization strategy to draw pictures, usually silly pictures, that link words spelled with the same pattern can be a strong association process that children typically enjoy.

What are the advantages of analyzing word elements for cuing spelling?

The word structure analysis process is very beneficial for children at later levels. It provides structure for spelling long words, cues meaning, teaches word elements, and makes polysyllabic words not the monsters they seem to be. Knowing how root words can be expanded to become polysyllabic words significantly enhances spelling, but only if that spelling takes into account both the auditory and visual elements in the words because children with dyslexia have a great tendency to drop syllables. Although the schwa phenomenon is more problematic in spelling lists for older children, fortunately, they have stronger short-term retention and retrieval from long-term storage to use in their analysis. Even with all these strategies, spelling the schwa phoneme is an obstinate problem that even good spellers face.

Merger of Phonology and Syntax Rules

How does the native speaker learn merged rules?

At certain points, pronunciation, syntax rules, and spelling patterns merge and are also rule bound. Most of these mergers are recognized automatically by the native speaker and do not require delineation of their relationships. But sometimes they confuse children who do not discriminate sounds well and have unstable understanding of morphologic rules. Some patterns that occur often and consistently can be pointed out didactically to help the decision-making process for phoneme-grapheme correspondence. For more complex and inconsistent patterns, giving more attention to practice and less to explaining the rules can habituate the rule more naturally. The following explanations for some of these patterns are available for reference if needed, but if a child does not have difficulty with a set of rules, they do not need to be pointed out. The native-speaker-process often clarifies better than didactic presentation to achieve competency.

Regular Past-Tense Pronunciation of /-ed/

How is past-tense pronunciation tied to preceding phonemes?

An example of a phonologic and morphologic rule merger is found in past tense /-ed/ pronunciation, as found in Table 10–1. Regular past-tense verbs all end with /-ed/, but if the last sound in the verb ends in a voiceless phoneme, the /-ed/ is pronounced [t]; if the word ends with a voiced phoneme, it is pronounced [d]; and if the word ends in [t] or [d], it is pronounced [əd]. Rules such as these can be helpful but can also become too difficult to use because they are dependent on retention and recall. Note that the morphologic /-ed/ rules will come up in the [d] and [t] spelling lists in the Neurolinguistic Approach to Reading (NAR). These patterns can be delineated at that point if the child shows confusion about the morphologic rules.

Plural, Possessive, and Third-Person Singular Word Endings

What are plural, possessive, and third-person singular morphologic rules for pronunciation and spelling?

The phonemes /s/ and /z/ also carry significant syntactic rules in final word position, in addition to their phonologic features. They designate plural, possessive, and third-person singular morphologic forms. The spelling rules for these syntactic forms are fairly consistent, but they typically confuse children learning to read and spell. Some may be able to process these rules, and others may need the information reduced and simplified

Table 10–1. Regular Past Tense Pronunciation of /-ed/

Word Ending in	Pronounced	Spelled /-ed/
Voiceless phoneme	[t]	kissed
voiced phoneme	[d]	stayed
[t] or [d]	[əd]	waited, ended

by, for instance, postponing the possessive rules. But, depending on the age, memory ability, and cognitive level of the child, three rules can be presented explicitly as follows:

1. If the sound immediately preceding /s/ is voiced, the plural, possessive, and third-person singular /s/ is pronounced [z], for example, *two cars, Mary's ball,* and *wags his tail.*
2. If the sound immediately preceding /s/ is unvoiced, the plural, possessive, and third-person singular /s/ is pronounced [s], for example, *two cakes, cat's whiskers,* and *he hits.*
3. If the sound immediately preceding the /s/ and /es/ markers for plural, possessive, and third-person singular is a sibilant sound, the /s/ and /es/ are pronounced [əz]. Plural and third-person singular forms are spelled /es/, but the possessive spelling form is /'s/, even though all forms are pronounced the same, for example, 2 *busses, bus's,* and *he kisses.*

Plural Noun Endings

How does voicing affect pronunciation of plural nouns?

The clinician must have clear knowledge of these patterns (and the exceptions such as irregular plural and irregular past-tense forms) to guide spelling. A child will often perceive the correct pronunciation of, for instance, *puppies* or *boxes* and attempt to spell the words phonetically as *puppyz or boxez.* These may be good times to introduce the concept of [s], [z], and [əz] options for letter coding with /s/, /es/, and /ies/ at the end of words while praising their excellent phonologic perception in hearing the [z] (Table 10–2).

When should the "exception to the rule" patterns be introduced?

Other morphologic rules governing pluralization that affect spelling can be introduced at readiness levels, and these spelling patterns can be postponed until spelling is progressing well. The rules are more complicated and less consistent. One of the problems children with dyslexia have when some of these rules are introduced in language arts instruction is that they are overwhelmed by the coding task, and the syntax information is never absorbed, even when repeated over several years. If children have difficulty with cause-effect reasoning, they will have difficulty because

Table 10–2. Plural Noun Endings

Word Ending in	Example Word	Change Spelling to	Pronounce
voiced C*	hug	/hugs/	[hʌgz]
unvoiced C*	hat	/hats/	[hæts]
voiced or unvoiced sibilant [s,ʃ,ʒ,tʃ,dʒ]	judge church	/judges/ /churches/	[dʒʌdʒəz] [tʃɝtʃəz]

Note. C* = Consonant.

the rules use *if-then* wording, sometimes application of exclusion reasoning with *except for*, and other times inclusion reasoning that uses *or*. Recall issues are often caused by poor storage due to not understanding the language that explained the rule in the first place, and that language weakness accounts for the errors, not forgetting the rule; it was never learned. As important as learning these rules early is to avoid reinforcement of incorrect understanding and spelling that confuse recall, caution is indicated because information is also not stored well if it is not within the zone of proximal development (ZPD), that is, beyond the child's ability level.

Plural Words Ending in /y/ and /ey/

How are nouns ending in [i] and [aɪ] made plural in spelling?

Plural patterns are unique for regular nouns that end with the phoneme pronounced [aɪ] or [i] and spelled with /y/, for example, *sky* and *bunny* as seen in Table 10–3. The word is made plural by saying [z] after the vowel, and the spelling is adjusted by changing /y/ to /i/ and adding /es/, for example, *skies* and *bunnies*. Words ending with the phoneme [i] that are spelled with a vowel before the /y/ are made plural by saying [z] after the vowel and simply adding /s/, for example, *attorneys* and *monkeys*.

Plural Words Ending in /f/

What are the exceptions for making words ending in [f] plural?

Pluralization of words ending in [f] (spelled /f/, /ff/, /ffe/, and /fe/) represents significant exception patterns, as seen in Table 10–4. Most nouns ending in [f] (spelled /f/, /ff/, /ffe/, or /fe/) are made plural by simply adding /s/ (spelled and pronounced [s]), for example, *roofs*, *chiefs*,

Table 10–3. Plural Words Ending in /y/ and /ey/

Word Ending in	Example Word	Change Spelling to	Pronounce
[i]: C+ /y/	puppy	/puppies/	[pʌpiz̲]
[i]: V*+/y/	monkey	/monkeys/	[mʌŋkiz̲]
[aɪ]: C+ /y/	supply	/supplies/	[səplaɪz]

*Note. V** = Vowel.

Table 10–4. Plural Words Ending in /f/

Singular Spelled	Plural Spelled	Plural Pronounced	Example Words
[f]: /f/, /ff/, ffe/, /ph/	/fs/, ffs/, /ffes/, /phs/	[fs]	*roofs, cliffs, giraffes, graphs*
[f]: /f/, /fe/	/ves/	[vz]	*wolves, shelves, knives, leaves*

giraffes, and *cliffs*. The irregular pattern in some words is that the ending [f] phoneme (spelled /f/, /ff/, or /fe/) is changed to [v] (spelled /ve/) plus [z] and is spelled /ves/, for example, *knives, wives, leaves, lives, slaves, selves, halves, calves*, and *shelves*. A few words can be said/spelled using either the regular or irregular pattern, for instance, *hoofs/hooves, scarfs/scarves*, and *dwarfs/dwarves*. At this level of irregularity, the clinician should be in touch with syntax evolution by consulting a current edition of an unabridged dictionary. These types of rules can change from one edition to the next as certain spellings become commonplace. Many children with dyslexia are very intelligent, and as their written language skills improve toward high school, they may have need of these rules more than might be expected. However, this level of orthographic rule learning is probably not necessary for younger children with dyslexia because learning and remembering the rules is beyond their skill level, and the spelling patterns are so peculiar that any errors could be accepted as almost normal at their age level. For those older students who would benefit from meta-analysis of phonologic patterns as they merge to encompass morphology, the patterns can be introduced.

Plural Words Ending in /o/

How are words ending in /o/ made plural?

Nouns ending in /o/ also have special pluralization rules, as seen in Table 10–5. If the /o/ is preceded by a vowel, the end sound is pronounced [z] and spelled /s/. If the word ends in a consonant plus /o/, the end sound is also pronounced [z] but spelled /es/. A small group of words of Italian origin, usually musical terms, do not follow the consonant plus /o/ rule and just add /s/ for plural. Again, these orthographic rules may overchallenge memory skills, and time would be better spent working on other vulnerable skills.

Root Word Change for Plural Nouns and Nouns Used Only in Plural

What nouns do not show plural with suffixes?

Although most plural spelling patterns are based on phonologic features of the final phoneme in the word, as is true for other morphologic markers such as past tense, the plural form of some nouns is not based on final phonemes. Instead, these words are made plural by changing phonemes

Table 10–5. Plural Words Ending in /o/

Singular Spelled	Plural Spelled	Plural Pronounced	Example Words
vowel + /o/	/os/	[oz] or [oʊz]	*studios, radios, stereos, zeros*
consonant +/o/	/oes/	[oz] or [oʊz]	*potatoes, tomatoes, heroes*
Italian music words	/os/	[oz] or [oʊz]	*pianos, solos, sopranos*

within the word, usually the vowel, for example, *louse/lice* and *foot/feet*. A few words use plural rules from other languages, for example, *crisis/crises*, *beau/beaux*, or *index/indices*. Other plural forms are nouns that represent pairs, for instance, *scissors* or *pants*, in which the word has only a plural form and takes a plural verb.

Zero Plural Nouns

What are zero plural nouns?

Zero plural nouns are spelled the same whether they are singular or plural, for example, *sheep*, *deer*, *bison*, *species*, *cod*, *Japanese*, *headquarters*, *moose*, *shrimp*, *fish*, *elk*, *quail*, *reindeer*, and *swine*, which adds another complication to the spelling patterns of SAE. The verb clarifies the plural versus singular status. Although only a few zero plural nouns exist in English, some are common words, so they often present as problematic to children and even adults.

Possessive Designation of Zero Plural Nouns

How is possessive designated in zero plural nouns?

The possessive form of these nouns also cannot be discerned auditorily because they are pronounced the same, and they are spelled the same, with the apostrophe before the /s/ distinguishing the possessive form, whether singular or plural. Table 10–6 demonstrates the relationship of plural and possessive forms in these nouns.

Possessive Designation of Regular Plural Nouns

Where is the apostrophe placed for possessives?

The apostrophe for possessive designation can be conveniently introduced in the context of these pronunciation and spelling rules if the child paraphrases and executes the reasoning process. Learning the patterns can be postponed but should be introduced as early as possible to avoid needing to unlearn patterns later on. These are patterns for which the elementary school child will be held increasingly accountable. Table 10–7 demonstrates pronunciation and placement of the apostrophe and /s/ in singular and plural forms.

Pronunciation of Third-Person Singular Verb

Does pronunciation of third-person singular, final /-s/ differ from plural final /-s/?

Verb morphologic forms include final /s/ designation of third-person singular with [s], [z], and [əz] pronunciations that are determined by preceding phonemes in the same manner that other /s/ endings are marked, as seen in Table 10–8.

Table 10–6. Possessive Designation of Zero Plural Nouns

Singular	sheep	*The sheep **is** eating.*
Plural	sheep	*The sheep **are** eating*
Singular and plural possessive	sheep's	*The sheep's barn is big.*

Table 10–7. Possessive Designation of Regular Plural Nouns

Word Ending in	Example Word	Change Spelling to	Pronounce [s, z, əz]
Voiced C or V singular	dog singer	/dog's/ tail /singer's/ voice	[dagz̲] [sɪŋɚz̲]
Unvoiced C singular	hat	/hat's/ feather	[hæts̲]
Voiced sibilant singular	box	/box's/ lid	[baksə̲z̲]
Voiced C or V plural	cars pillows	/cars'/ patterns /pillows'/ shapes	[karz̲] [pɪloʊz̲]
Unvoiced C plural	hats	/hats'/ feathers	[hæts̲]
Voiced sibilant plural	busses (or buses)	/buses'/ schedules	[bʌsə̲z̲]
Unvoiced sibilant plural	quizzes	/quizzes'/importance	[kwɪzə̲z̲]

Table 10–8. Pronunciation of Third-Person Singular Verb

Word Ending in	Example Word	Change Spelling to	Pronounce [s, z, əz]
Voiced C* or V*	run glow	He /runs/. It /glows/.	[z̲] [z̲]
Unvoiced C*	trot	She /trots/.	[s̲]
Voiced sibilant	challenge	He /challenges/.	[ə̲z̲]
Unvoiced sibilant	kiss	She /kisses/.	[ə̲z̲]

Note. C* = Consonant; V* = Vowel.

How are verbs ending in /y/ spelled?

Third-person singular verbs ending in /y/ follow the same pattern as plural nouns ending in /y/, for example, *flies*, *spies*, and *tries*. If delineating the understructure patterns represented in the previous charts and descriptions does not facilitate learning, the words can simply be learned based on the way the phonemes sound with corrections for errors. Sometimes the understructure clicks with didactic instruction, and sometimes it does not. Even if a child is capable of memorizing the rules, rotely memorized information is subject to more rapid deterioration than analyzed, understood, and repeatedly applied information. So if the rules are memorized, the child should be asked periodically to explain the rule as a way of testing its stability. Ongoing oral and written language samples are good ways to assess knowledge status and need for further explicit rule instruction about understructure.

Why is American English not totally phonetic?

These charts represent the vast majority of spelling and pronunciation patterns for plural, possessive, and third-person forms, but many exceptions to spelling patterns occur. They often trace back to an evolutional history of word formation that only a linguist might care to decipher. With such varied sets of spelling options, often determined by oral pronunciation patterns that have changed over time while spelling has not, the question arises, *Why not adopt a phonetic language in which one grapheme represents one phoneme and one phoneme represents one grapheme?* Various local and national campaigns over the decades have proposed phonetic spelling of English, even as bills to Congress, but they have failed to gain general support and have little hope of ever accomplishing the goal. The reasons the task seems impossible are many and include the fact that, in general, English is already spelled somewhat phonetically, and American English has roots in Germanic language and British English, both of which have roots in many other languages, primarily due to many postwar population assimilations over the centuries, and those patterns have been carried forward. Changing spellings would require massive reinstruction in the United States, as well as reinstruction for international users of English, limiting commerce and international communication. The confusion caused by thousands of SAE homophones alone would require careful evaluation.

Why does weak pronunciation block ESOL learners?

Those learning English as a second or other language (ESOL) have particular difficulty with the pronunciation and orthographic rules governing possessive, plural, and third-person singular word forms and other suffix patterns on both a phonologic and morphologic basis. Small units of instruction with much opportunity for practice can clarify some of these inconsistencies. ELLs, in general, should gain as much mastery of English phoneme pronunciation patterns as possible because their inability to understand rules for syntax are often due to their not having perceived and pronounced the phonemes that change case, number, gender, and tense forms. If their native language does not use phonemes in certain word positions, has slightly different phonemes, or does not have the phoneme at all, they find it difficult to understand the morphologic rules imposed on those phonemes. Each step of syntax learning becomes blocked by the pronunciation difficulty, which, in turn, makes listening, reading, talking, and writing difficult.

What is the goal of remediation?

The reality of SAE spelling is that it is inconsistent, and yet correct spelling is a badge of literacy. Those whose writing does not reflect a certain level of spelling accuracy are, rightly or wrongly, judged to be lacking in some combination of education, intelligence, parental guidance, cultural compliance, or other fairly negative trait. Moats (2005–2006, p. 14) points out that in 2005, the National Commission on Writing for America's Families, Schools, and Colleges reported that 80% of the time, an employment application with poor writing and spelling is doomed. This, obviously, is unfair to individuals with dyslexia. The mission of remediation is to close the gap between obvious capabilities and weak display of literacy so that individuals with dyslexia are not faced with these negative judgments. But society should take a broader look at literacy. A "word" person (someone literate in reading and writing words) who cannot design a skyscraper, as a person with dyslexia might, is not considered lacking ability. Why should a person with dyslexia be judged for his set of strengths and weaknesses and the good reader/writer not considered handicapped for not being able to build the skyscraper? The analogy is not to belittle good reader/writers. Obviously, all such judgments should cease, but the point is that society and, certainly, the education process should treat these children with less negative judgment. The stories adults with dyslexia tell of their childhood reading and spelling misery should be taken more seriously. Many grown men and women have shed tears at their child's postassessment conferences as they share their own painful experiences, explaining that they, too, were children with dyslexia. What does not kill them does not make them stronger; the wounds still hurt.

Why is remediation important, even if it does not "cure" dyslexia?

Some children will not be able to totally conquer the difficult pronunciation and spelling patterns of SAE, but attempts to remediate the condition are well worth the effort because routinely they make significant gains. Consider those children with dyslexia who have unsuspected journalistic skills beneath their weak spelling and reading. Some of them write and use novel wording for unique ideas, have a bent toward poetry, present imagery that is palpable in print, and write unimaginable storylines. They may never gain total mastery of spelling and reading, but technology (and editors) will help them if they can get to a reasonable literacy level to interface with the "word" world. Even for individuals with dyslexia without such talents, gains in spelling typically release other reading and writing competencies.

Reference

Moats, L. C. (2005–2006). How spelling supports reading: And why it is more regular and predicable than you think. *American Educator*, pp. 12–43. Retrieved from http://www.aft.org/pdfs/americaneducator/winter0506/Moats.pdf

11

Logistics

Using the Block-by-Block Spelling Charts

How should the spelling forms or whiteboards be used?

The writing arrangement for this stage of the Neurolinguistic Approach to Reading (NAR) program can be adjusted to the motor skills of the child. For handwriting, a lined dry-erase board or lined paper can be used for writing the letters. NAR uses a sound-by-sound spelling grid that can be found in Appendix E, Table E–1. The child says the word, says the first phoneme in the word, writes the letter to represent that phoneme in the first space, says the next phoneme, writes the letter to represent that phoneme in the second space, and so on, until the word is spelled. The child uses only one box for each phoneme, whether the phoneme is spelled with one or four letters. The child then says the word again and spells it as a nonsegmented word. Any errors should be corrected before the second spelling and discussed for the problems involved. The error may represent another spelling of the target phoneme, misspelling of a nontarget phoneme, letter reversal, or other inaccuracies. If patterns become apparent, exercises can be added that clarify the error issue for the child. Frequent misspelling of a nontarget phoneme often suggests that phoneme as a good candidate for the next targeted phoneme.

What is the advantage of syllable spelling?

Another form is used for syllable spelling and is also found in Appendix E, Table E–2. As the child faces an increasing number of polysyllabic words at about third grade, it can be helpful to spell words syllable by syllable. Syllable stress represents the beat (loudness and duration) and pitch changes in ongoing words. Awareness and designation of stress patterns in words provides a clearer understanding of the beat and melody pattern on which words are said. For children having difficulty detecting stress, a useful technique is found in Harold Edwards's workbook that accompanies his text (Edwards & Gregg, 1992). Where the child breaks the syllable does not matter for NAR syllable spelling, only that the syllable represents one beat. Depending on the accuracy of syllable spelling, the child may need to also add spelling sound by sound within the syllable. Eventually, the child will be encouraged to spell syllable by syllable to avoid the syllable deletion problem that many children have and children with dyslexia have more often. Adults helping prepare the child for the school spelling test should try to reinforce this approach to spelling. The adult should say the spelling words from school and have the child tap the number of syllables, then say each sound in the syllable before writing the letter. They should not look at words and try to memorize their visual configuration.

What is a reasonable mastery level?

Reviewing and testing missed words should continue until only a few errors are made. Although the percentage of wanted accuracy may vary, depending on the child, higher percentages, such and 90% or 95%, yield better storage and recall of patterns from long-term stored memory, but they may be unrealistic for some children. It may be necessary to accept accuracy rates as low as 70%. But it should be kept in mind that the original score was probably much lower, so the child has still gained skill. Also, one spelling list is not the last time the child will visit words like the ones on that list, and as different lists for different phonemes have similar patterns, the child will gain tools to sort patterns, and so will that child's language acquisition device (LAD).

What is the difference between sorting letters for phonemes and sorting phonemes for letters?

Some children benefit from considering the fact that the same letter pattern can "say" different sounds, for instance, /bow and arrow/ compared to /brown/. This type of distinction is used routinely in Phono-Graphix exercises, but NAR does not stress this strategy unless a child is particularly confused by the phenomenon of same letters saying different phonemes. NAR should have made it perfectly clear that, of course, same letters say different things, but this type of exercise is not reinforcing the desired skill for coding with letters. Once spelling sounds with letters is strong, considering different phonemes that letters say may be interesting for some children, but the goal is for the child to always consider phoneme patterns first because, if anything, usually visual pattern recognition is already too much the leader in the coding process.

How can semantic and syntactic elements be incorporated into the process?

A routine strategy used in NAR is to add semantic and syntactic elements to the storage process by producing sentences using the target word and writing a little story using several of the words from the phoneme list. Children particularly appreciate this strategy and typically keep a notebook of all their work to verify their efforts. A checklist at the front of these exercises keeps track of the phoneme lists that they have finished (Appendix E, Table E–3). A younger child can dictate a story with words on the list while the clinician types exactly what the child says. Great humor, unexpected thoughts, and creativity emerge at this level, and the children feel the strength of their ideas. The entire spelling-reading cycle is complete. The child heard the words, spelled the sounds in the words, wrote sentences, wrote the story, read the word list, read the sentences, and read the story. The fact that the child wrote the story makes the child invested, usually proud in reading it, and remembers much of what was written, which facilitates fluency. If children dictate to expedite the process, in other exercises they can write their own sentences and stories.

The Grade-Level Spelling Lists in Appendix J

How should sight words be selected for memorizing?

Educators often refer to high-frequency sight words, meaning that some words are used so often that they should be sight-read. NAR generally discourages this approach to reading because it teaches the opposite reasoning process needed for coding and encourages children to depend on sight-reading instead of auditory reading. However, a few words have such atypical spellings and are used infrequently enough that making the child learn a spelling option for a sound that represents just a few words in the language may be unjustified. These words, such as *two* in which [u] is spelled /wo/, need to be assessed by the adult to determine the best way to teach the child. Sometimes these outlier spelling patterns for words are actually easier for children to remember because they are so unique, or the word is used so often and is so short that they are able to store the spelling pattern. The language has a limited number of such words, and whenever these words occur in the NAR grade-level lists, they are placed in the "Sight Vocabulary" box at the end of the grade section. Most sight word lists teachers use are comprised of words that actually have frequently used phoneme coding patterns. The wall words or other strategies that encourage memorizing how words look can be problematic for children with dyslexia until they reach a certain competency level—and then they probably will not need the visual reference. As Orton (1937/1999, pp. 199–206) pointed out in "The 'Sight Reading' Method of Teaching Reading as a Source of Reading Disability" section of his book, *Reading, Writing, and Speech Problems in Children*, sight-reading not only prevents academic learning but also has potential for far-reaching damage to the emotional life for those children struggling with reading acquisition. Even on a limited basis, not a full-blown instructional reading method, the strategy is counterproductive.

What is a primary objective when working on higher grade-level sets?

Advanced-level vocabulary in older lists displays more atypical spelling options because words at this level have maintained the spelling patterns of phonemes of the language from which the word was derived, and often prefixes and suffixes modify the root words. Children and adults working with the high school-level lists are mastering their dyslexia. They are typically trying to bring their reading fluency in line with their cognitive level. They are ready to tackle the orthographic phoneme rules such as those that cue for double consonants, specific foreign language forms, and more extensive consideration of root, suffix, and prefix rules. At this level, the coding process is highly intertwined with semantic and syntactic processes, and phonology is only part of what needs to be addressed to advance written language skills. As a result, lists above sixth-grade level represent a more complex linguistic task and require a different type of motivation.

What is the broken or silent /e/ rule?

One orthographic letter spelling pattern that is useful and should not overtax the vulnerable memory of children with dyslexia is the "broken /e/" or "silent /e/" rule for spelling vowels. The child can reconsider any word ending in /e/ to determine if the silent /e/ rule applies, as in *care*, in which the /e/ "makes the /a/ say its name." The silent /e/ orthographic rule is introduced cautiously in NAR, with assessment of the child's ability to remember and use such an inconsistent rule. This spelling option allows "long" vowels to say their names, that is, [eɪ], [i], [aɪ], [oʊ], and [ju]. In NAR, this pattern is designated by the letter that says its name, that is, /a/, /e/, /i/, /o/, and /u/, followed by a blank line, and then the letter /e/, that is, /a_e/, /e_e/, /i_e/, /o_e/, and /u_e/. One of several (but not all) consonant letters can be put in the blank, and the pronunciation will match the vowel's letter name. Because this pattern occurs often for five different vowels, it should be considered early as a spelling option, but the exceptions, such as /have/, /love/, and /give/ that trigger yet other orthographic rule patterns, can be confusing for children.

When can most of the spelling options be found in children's reading material?

The kindergarten list typically has one or two spelling options for consonants and two or three options for vowels. By second grade, those spelling options have nearly doubled, but few other options are added after that, perhaps 20 over the next 10 years. Those additions are often for words from other languages that are commonly used in SAE to which older children are exposed, for example, [oʊ] being spelled /eau/ in the word *beau*. Certain vowel phonemes are particularly problematic because of the large number of spelling options, for instance, the phoneme [eɪ], which has nine spelling options, and [aɪ], which has seven options in the fifth-grade list. Although these options can be challenging, children identify similar patterns across vowel spellings and carry out association exercises to help their visual recall, which makes the task easier, and they take pride in selecting the correct pattern in a word they are spelling.

Why do some children with weak visual recall still sight-read?

Sometimes children store a spelling pattern based on incorrect visual processing or weak storage, and the pattern can be detected, for instance, when the child spells *yewllo* for *yellow* or *popele* for *people*. They remember something about the letter pattern that is correct, but they are not using phonologic perception to order the sequence. The tendency to store words visually without accurate representation of phonemes and phoneme sequence at elementary ages can make remediation more difficult, especially when, in addition to weak auditory processing, visual processing is weak. Why some children with weak visual perception depend on visual recall for spelling and reading remains an unresolved question, perhaps because, though weak, the visual skills are better than the auditory ones. Another pattern seen is the child who shows weak phonologic perception, gains excellent discrimination in therapy, but then goes on to spell words totally phonologically, but incorrectly, and only masters correct spelling with inordinate clinical effort. When this occurs, suspicion is raised about possible phonics exposure that taught only the most frequently used letters that code phonemes.

Analyzing errors often gives insight to cause, allowing refocus of effort on areas of confusion. In Appendix F, Table F–1 provides a miscue analysis, with a blank log to record error types in Table F–2.

What is the difference between balanced and equal sensory involvement?

The focus for speller-readers should be on balancing auditory and visual retrieval, starting with improved understanding of the mouth and its relationship to making sounds, phonologic perception, and then memorization of correct options for spelling phonemes in words using phonologic cue, visual recall, semantic associations, pattern groupings, and other strategies. Balance does not mean equal auditory and visual coding; it means auditory lead gained through auditory training and phoneme coding. NAR provides an awakening of skills necessary to spell and read. Children can depend on what they hear to unlock the fundamentals of an unfamiliar word for spelling and then consider their options for spelling the sounds. Then they solidify the correct choice with visual confirmation.

How do spelling patterns generalize?

Table 11–1 summarizes spelling options at a fifth-grade level, which represents a functional level for reading many newspapers, although newspaper reading levels vary and can reach college level for some reports and sections. As higher level vocabulary is introduced beyond mastered fifth-grade spelling levels, many patterns for spelling choices have been solidified by practice with earlier sets, so the coding process becomes more automatic, based on the theory that neurobiologic categorization has improved and takes over much of the sorting process.

What are some more phonetic phonemes at the fifth-grade level?

Sample charts prepared by different children are presented in Appendix G, Table G–1 for a kindergarten child, G–2 first grade, G–3 second grade, G–4 third grade, G–5 fourth grade, G–6 through G–8 fifth grade, and G–9 seventh grade.

Table 11–1. Fifth-Grade Spelling Options for Consonant Phonemes

Phoneme	Spelling Options	Phoneme	Spelling Options
[p] *pie*	p, pp	[r] *run/car*	r, rr, wr
[b] *boy*	b, bb	[l] *like*	l, ll, le, el, al, ell, il
[t] *to*	t, ed. tt	[j] *yellow*	y, i
[d] *dog*	d, dd, ed, t, tt	[ju] *bugle*	u_e, ew, ue, eau, u
[k] *cat*	k,c, ck, ch, cc, qu	[i] *eat*	ee, e, y, e_e, ie, ea, ey, i, ei
[kw] *quack*	qu	[ɪ] *it*	i, y, ui
[ks] *box*	x, xc, cc	[ɛ] *egg*	ai, e, ea, ie, a
[g] *go*	g, gg, gh	[æ] *at*	a
[gz] *exhibit*	x	[ɑ]/[ɔ] *not/paw*	a, o, al, ough, au, aw, u, oa
[s] *so*	s, se, c, ss, ce, sc	[ʊ] *book*	oo, oul, u
[z] *zoo*	s, z, se, zz, ze, ss	[u] *shoe*	ue, o, ew, oo, oe, ui, ou, u_e, u
[ʃ] *she*	sh, ti, s, ch, ci, c, si, ss	[ɝ]/[ɚ] *further*	or, ir, ear, er, ur, urr, irr, ure, ar, our
[ʒ] *beige*	si, s	[ʌ]/[ə] *above*	u, a, o, e, i, o_e, ou, oo
[tʃ] *chop*	ch, tch, t	[eɪ]/[e] *vacation*	ay, a, a_e, ai, ey, eigh, aigh, ei
[dʒ] *joy*	j, dge, ge, g, dg	[oʊ]/[o] *hobo*	o, o_e, oa, ow, ou, oe
[f] *fun*	f, ff, gh, ph	[aʊ] *out*	ou, ow, ough
[v] *vine*	ve, v	[aɪ] *eye*	y, i, i_e, igh, ie, ey, ai
[θ] *think*	th	[oɪ] *boy*	oy, oi
[ð] *that*	th	[ʔ] *kitten*	tt
[h] *hat*	h		
[m] *my*	m, mb, mm, mn		
[n] *no*	n, nn, kn, gn		
[ŋ] *ring*	ng, n		
[w] *we/what*	w, wh		

References

Edwards, H. T., & Gregg, A. L. (1992). *Applied phonetics workbook: A systematic approach to phonetic transcription*. San Diego, CA: Singular.

Orton, S. T. (1999). *Reading, writing, and speech problems in children and selected papers*. Baltimore, MD: The International Dyslexia Association. (Original work published 1937)

12

Summary

As the information presented in the Neurolinguistic Approach to Reading (NAR) is considered in retrospect, the words of Chomsky come to mind. How could a small baby sort the infinite number of possible phonemes to come to recognize and say words? And how could the progression of that knowledge balloon in the first few years of life to encompass the complex rule systems of syntax, morphology, semantics, pragmatics, and prosody? NAR has presented but a sampling of how the brain theoretically negotiates language. Contradictory, modified, newer, reconsidered, or just different theories about language and language disorder exist or may emerge. Whether the process involves a language acquisition device or a generalized, mega, computer-like sorting process, the outcome is awe-inspiring. While we grapple with the complexity of language in our efforts to sort its rules, it is important that what we know about how the brain processes and expresses language form and content be applied to individuals with dyslexia. We also need to be ready to make revisions in our thinking as new research and clinical insights give us clearer understanding of the ways that auditory and visual coding skills integrate in written language. Meanwhile, we must apply our theories.

APPENDIX A

Alphabet Charts

Table A–1. Random Lowercase Alphabet

d	z	n	a	q	l	j
r	c	x	k	f	y	o
e	s	t	m	i	p	h
b	g	u	w	v		

The child says the letter name for each letter, and the response is recorded on the second column of the tracking chart.

Table A–2. Lowercase Letters Without Serifs

a	b	c	d	e	f	g
h	i	j	k	l	m	n
o	p	q	r	s	t	u
v	w	x	y	z		

Sometimes the serifs on letters confuse preschool and kindergarten children, and using letters without serifs is helpful.

Table A–3. Uppercase Alphabet in Order

A	B	C	D	E	F	G
H	I	J	K	L	M	N
O	P	Q	R	S	T	U
V	W	X	Y	Z		

Although lowercase letters are almost always used with the Neurolinguistic Approach to Reading (NAR), uppercase forms may be helpful for some children. A few may be added as needed, and when ready, uppercase letters must be systematically learned as well. The adult can point to letters randomly and have the child name them, with results recorded on the tracking form.

Table A–4. Lowercase Alphabet in Order

a	b	c	d	e	f	g
h	i	j	k	l	m	n
o	p	q	r	s	t	u
v	w	x	y	z		

Table A–5. Alphabet Tracking Chart for Mastery

Point to named letters						Name letters						Write letters					
i						i						i					
a						a						a					
q						q						q					
r						r						r					
b						b						b					
k						k						k					
w						w						w					
x						x						x					
h						h						h					
z						z						z					
g						g						g					
f						f						f					
l						l						l					
e						e						e					
d						d						d					
j						j						j					
n						n						n					
m						m						m					
c						c						c					
p						p						p					
s						s						s					
o						o						o					
t						t						t					
u						u						u					
v						v						v					
y						y						y					
Total:						Total:						Total:					

This chart requires total mastery of all three tasks before any phoneme-letter matching exercises are undertaken. If the percentage or number correct is tracked on a bar chart, the child typically accepts the challenge and gains alphabet mastery fairly easily. Errors can be documented by noting the substitutions or *I don't know* responses of the child. Attempts to replicate the way in which the child wrote the letter can be noted in the *Write letters* column. The chart allows five trials and can be duplicated as many times as necessary, with copies of all the hard work that includes a sticker, kept in the notebook to allow children to feel the importance of their efforts. Stickers help.

Table A–6. Reversal Exercise for /b/ and /d/

d	b	b	d	d	b	d
b	d	b	d	b	b	d
b	b	b	d	b	b	d
d	b	d	b	d	d	b

After discussion and activities to clarify names used to describe directionality, a child having reversal problems can be asked to perform tasks such as putting red chips on all the /b/ letters and yellow chips on all the /d/ letters, or if children have sufficient ordering and counting skills, they can be asked to point to the second /b/ in the first row, third /d/ in the fourth row, and so on.

Table A–7. Reversal Exercise for /b/, /g/, /p/, and /q/

b	g	p	b	q	q	q
b	b	b	d	p	d	b
g	q	b	d	q	d	b
p	g	d	g	b	q	p

For children having confusion with these letters, they can be asked to name each letter until all are recited correctly, or the same strategies used for /d/ and b reversal can be applied.

Table A–8. Digraph Discrimination Exercise

sh	t̲h̲	ch	th	ch	sh	t̲h̲	ch	th
ch	sh	t̲h̲	ch	th	sh	ch	t̲h̲	sh
sh	ch	th	ch	sh	t̲h̲	ch	sh	th
ch	sh	t̲h̲	th	ch	sh	ch	th	t̲h̲

Beyond the alphabet-learning stage, as children are learning to associate phonemes with visual symbols, some are confused by digraph symbols in which two letters make one phoneme. This can be clarified over time by naming each one on the chart above until all are said correctly. Digraph /ch/ is [tʃ], /sh/ is [ʃ], /t̲h̲/ is [ð], and /th/ is [θ].

APPENDIX B

Consonant and Vowel Distinctive Feature Charts

Children will vary in their need to have distinctive features emphasized, but a brief presentation of all phonemes and phoneme features should at least be given to document how quickly and accurately the child recognizes and explains phoneme features. Parts or all of the charts can be used as exercises or as a reference. They represent feature designations by Chomsky and Halle (1968), Kantner and West (1960), Bernthal and Bankson (1993), and Edwards (2002).

These charts can be used in a variety of ways. They can be broken into smaller focus points, for instance, nasal sounds to clarify phoneme differences within the feature of nasality. They can be used as references during therapy when two phonemes are not distinguished for their differences based on how they sound. After orienting the child to what the charts represent, the charts can be used to assess gains by having the child complete the chart before being presented extended clinical analysis. The point to be made about the phonemes is that each one has its own unique pattern. A few examples are provided for targeting nasality and glides as well as charts for all consonants and all vowels. A feature is described as either present (+) or absent (−).

Table B–1. Nasality Distinctive Feature Focus

Phoneme	Stop	Front	Labial	Sibilant	Nasal	Voiced	Dental	Glide	Glottal	Fricative
my [m]										
no [n]										
ring [ŋ]										

Phoneme	Stop	Front	Labial	Sibilant	Nasal	Voiced	Dental	Glide	Glottal	Fricative
my [m]	–	+	+	–	+	+	–	–	–	–
no [n]	–	+	–	–	+	+	–	–	–	–
ring [ŋ]	–	–	–	–	+	+	–	–	–	–

Table B–2. Glide Distinctive Feature Focus

Phoneme	Stop	Front	Labial	Sibilant	Nasal	Voiced	Dental	Glide	Glottal	Fricative
you [j]										
run [r]										
like [l]										
we [w]										

Phoneme	Stop	Front	Labial	Sibilant	Nasal	Voiced	Dental	Glide	Glottal	Fricative
you [j]	–	+	–	–	–	+	–	+	–	–
run [r]	–	+	–	–	–	+	–	+	–	–
like [l]	–	+	–	–	–	+	–	+	–	–
we [w]	–	–	+	–	–	+	–	–	–	–

Table B–3. All Consonant and Glide Distinctive Features

Phoneme	Stop	Front	Labial	Sibilant	Nasal	Voiced	Dental	Glide	Glottal	Fricative
boy [b]										
that [ð]										
joy [dʒ]										
fun [f]										
go [g]										
you [j]										
cat [k]										
like [l]										
my [m]										
no [n]										
ring [ŋ]										
pie [p]										
run [r]										
so [s]										
to [t]										
chop [tʃ]										
vine [v]										
we [w]										
zoo [z]										
measure [ʒ]										
think [θ]										
she [ʃ]										
dog [d]										
hat [h]										

continues

Table B–3. *continued*

Phoneme	Stop	Front	Labial	Sibilant	Nasal	Voiced	Dental	Glide	Glottal	Fricative
boy [b]	+	+	+	–	–	+	–	–	–	–
that [ð]	–	+	–	–	–	+	+	–	–	+
joy [dʒ]	+	+	–	+	–	+	–	–	–	–
fun [f]	–	+	+	–	–	–	+	–	–	+
go [g]	+	–	–	–	–	+	–	–	–	–
you [j]	–	+	–	–	–	+	–	+	–	–
cat [k]	+	–	–	–	–	–	–	–	–	–
like [l]	–	+	–	–	–	+	–	+	–	–
my [m]	–	+	+	–	+	+	–	–	–	–
no [n]	–	+	–	–	+	+	–	–	–	–
ring [ŋ]	–	–	–	–	+	+	–	–	–	–
pie [p]	+	+	+	–	–	–	–	–	–	–
run [r]	–	+	–	–	–	+	–	+	–	–
so [s]	–	+	–	+	–	–	–	–	–	–
to [t]	+	+	–	–	–	–	–	–	–	–
chop [tʃ]	+	+	–	+	–	–	–	–	–	–
vine [v]	–	+	+	–	–	+	+	–	–	+
we [w]	–	–	+	–	–	+	–	–	–	–
zoo [z]	–	+	–	+	+	–	–	–	–	–
measure [ʒ]	–	–	–	+	–	+	–	–	–	–
think [θ]	–	+	–	–	–	–	+	–	–	+
she [ʃ]	–	–	–	+	–	–	–	–	–	–
dog [d]	+	+	–	–	–	+	–	–	–	–
hat [h]	–	–	–	–	–	–	–	–	+	–

Slight modifications to Chomsky and Halle's categorizations of distinctive features have been made to facilitate understanding of a binary classification system, for instance, *high/low, front/back,* and *rounded/spread*. All vowels are voiced (phonated at the vocal folds), sonorant (loud), and nonconsonantal (have less restricted airflow), so those features should be considered positive for all vowels. These charts represent the 14 vowels and six diphthongs of Standard American English (SAE). Discrimination of these phonemes is more difficult because of the relatively minute physical area in which the tongue makes them, never quite touching other parts of the mouth.

Table B–4. Vowel Distinctive Features

Phoneme		Front	Back	High	Low	Central	Long	Short	Rounded	Spread
eat	[i]									
it	[ɪ]									
vacation	[e]									
every	[ɛ]									
at	[æ]									
above	[ə]									
above	[ʌ]									
further	[ɝ]									
further	[ɚ]									
father	[ɑ]									
or	[ɔ]									
hobo	[o]									
book	[ʊ]									
shoe	[u]									
hobo	[oʊ]									
vacation	[eɪ]									
by	[aɪ]									
out	[aʊ]									
boy	[ɔɪ]									
unit	[ju]									

continues

Table B–4. *continued*

Phoneme		Front	Back	High	Low	Central	Long	Short	Rounded	Spread
eat	[i]	+	−	+	−	−	+	−	−	+
it	[ɪ]	+	−	+	−	−	−	+	−	+
vacation	[e]	+	−	−	−	−	+	−	−	−
every	[ɛ]	+	−	−	−	−	−	+	−	−
at	[æ]	+	−	−	+	−	−	+	−	−
above	[ə]	−	−	−	−	+	−	+	−	−
above	[ʌ]	−	−	−	−	+	+	−	−	−
further	[ɝ]	−	−	−	−	+	+	−	+	−
further	[ɚ]	−	−	−	−	+	−	+	+	−
father	[ɑ]	−	+	−	+	−	+	−	−	−
or	[ɔ]	−	+	−	−	−	−	+	+	−
hobo	[o]	−	+	−	−	−	+	−	+	−
book	[ʊ]	−	+	+	−	−	−	+	+	−
shoe	[u]	−	+	+	−	−	+	−	+	−
hobo	[oʊ]	−/−	+/+	−/+	−	−/−	−/+	+/−	+/+	−/−
vacation	[eɪ]	+/+	−/−	−/+	−/−	−/−	+/−	−/+	−/−	−/+
by	[aɪ]	+/+	−/−	−/+	+/−	−/−	−/−	+/+	−/−	−/+
out	[aʊ]	+/−	−/+	−/+	+/−	−/−	−/−	+/+	−/+	−/−
boy	[ɔɪ]	−/+	+/−	−/+	−/+	−/−	−/−	+/+	+/−	−/+
unit	[ju]	+/−	−/+	+/+	−/−	−/−	−/+	+/−	−/+	+/−

The information in these charts may be too technical for many children with dyslexia and adults to appreciate, but parts of the charts can be useful for different ages and ability levels. For instance, classification of a phoneme as long or short represents a different duration classification system than children are familiar with, and it might be expedient to bypass the long/short feature unless needed. An example of a situation in which clarification of duration might be helpful is the exceptional child who accurately hears two sounds in the diphonemes and wants to represent both, for instance, spelling *use* as *yuse*. The chart also serves the useful purpose of reminding the clinician of distinctive feature confusions the spelling and reading errors suggest so that those features can be built into exercises that clarify the uniqueness of involved phonemes. Two important messages these charts provide are that the child's neural mechanism is making all of these and many more distinctions and that breakdowns are usually linked events. As a result, discrimination breakdown can be more widespread than might seem to be the case.

References

Bernthal, J. E., & Bankson, N. W. (1993). *Articulation and phonological disorders* (3rd ed.). Englewood Cliffs, NJ: Prentice-Hall.

Chomsky, N., & Halle, M. (1968). *The sound pattern of English*. New York, NY: Harper & Row.

Edwards, H. T. (2002). *Applied phonetics: The sounds of American English* (3rd ed.). Clifton Park, NY: Delmar, Cengage Learning.

Kantner, C. E., & West, R. (1960). *Phonetics*. New York, NY: Harper & Brothers. (Original work published 1933)

APPENDIX C

Stage II Activities

During this stage of preparation for auditory discrimination, children are exposed to information that is readily understandable but has probably never been brought to their attention previously, and they are often pleased that they can answer many questions about how they produce speech. Information should be taught systematically, by phoneme and features, to be certain that significant clues for perception of each phoneme in the language are not neglected.

Using the grade-level spelling lists that will be selected for children to spell and read is a good resource when they are ready to code phonemes with letters. However, other resources can be added, including kits, manuals, close-up mouth photographs, anatomy pictures and charts, and many other strategies that the speech-language pathologist (SLP) routinely uses. How the information is grouped for presentation depends on the clinician's observation of the child's ability to retain information for short and long periods of time. Mirrors, nasal olives, tongue depressors, coffee stirrers, peanut butter and other taste-enhancing substances, pictures, diagrams, and other manipulatives can make these activities more interesting and meaningful than simple discussion.

Objectives include increasing sensitivity to what the mouth does to produce each phoneme in the language; associating that knowledge with what is heard; grouping phonemes that share distinctive features; being able to name targeted distinctive features of phonemes; and, particularly for children already reading and spelling, emphasizing features that spelling and reading error patterns have indicated are vulnerable.

The classes of phonemes in the Neurolinguistic Approach to Reading (NAR) program provide general descriptions of consonant and vowel distinctive features, but clinicians will need to elaborate on phonologic processes based on the child's perceptions and misperceptions. For instance, a child may readily detect the back/front feature in [k/t] but not in other [g/z]. Such fine distinctions may not be necessary, but back/front differentiation is an important auditory discrimination distinction that many children with central auditory processing disorder and dyslexia display, so learning to discriminate polar contrasts such as [g/z] should be helpful.

The vibration features of nasality and voicing are usually easily detected. Even though partial voicing and assimilation nasality occur, this level of differentiation is not necessary unless a child mishears nasality and voicing and misspells accordingly. General parameters of distinctive features are the focus. Subtle confusions such as labiodentalization of glides [r] and [w] often yield a /v/ spelling based on faulty perception and will need clarification. In working with a child, these confusions will be detected, especially in nonword activities that provide no opportunity for memorized spelling.

Voiced and unvoiced consonant cognates are usually a good starting point for feature discrimination because children usually can feel the voicing vibration and realize that they have mastered a critical binary feature involving 16 phonemes ([p:b, t:d, k:g, f:v, s:z, ʃ:ʒ, θ:ð, tʃ:dʒ]) that they never considered previously.

The vowel diagram is a useful tool to visualize tongue height in relation to the top of the mouth, back/front, and open/closed features. Because the three-dimensional area in which the tongue moves is very small, listening must be paired with miniscule movements between adjacent vowels to distinguish differences. For this reason, polar contrasting between [i], [ɑ], and [u] may be an easier starting point to demonstrate to the child that the tongue operates within a small space but with discrete locales for each vowel in the language. Phonemes closer to each other in proximity may sound very similar, but tactile and kinesthetic feedback can help confirm differences that are typically spelled differently. The index finger can be placed on the jaw as the high front to low front vowels are said to help distinguish vowels that are often perceptually confused, that is, [ɪ], [ɛ], and [æ].

Didactic information can be presented in many types of activities. Simple question/answer activities are efficient but may need an objective to sustain interest, for instance, answering 20 questions in a row correctly as tallied by a smiling versus frowning face or moving from a 70% to a 95% correct criterion for an older child. After the child demonstrates basic understanding, knowledge can be confirmed through further questioning.

Pictures of words can be used to identify designated phonemes with those phonemes then examined for their distinctive features. Minimal pair sets can be presented and the child asked what the phoneme or features were that made the words rhyme. The child can be asked to identify the one of six pictured words displayed on the table or the word for an object in the room that begins with an unvoiced hissing sound; a noisy, back popping sound; or a quiet, dental, friction sound. The task can be made more or less challenging by adjusting the field of choices and number of phonemes/distinctive features to be considered. Parents can extend these identification games in the car or at home, usually with an easier criterion such as, "Name a word that starts [or ends] with ___," letting the person who first names a certain number of words choose the next phoneme (not letter).

Mirror work verifies tactile and kinesthetic feedback for many features. The mirror is also useful for comparison of mouth structures to pictures or diagrams of targeted articulators. Shared mirror examination of mouth features is particularly enlightening, so large mirrors are needed along with small ones that can confirm nasal emission for nasal phonemes.

Fast naming of words beginning with a specific phoneme in a 1- or 2-minute period solidifies perception of feature quality and also improves retrieval from long-term stored memory.

A number of pictures can be displayed and the child asked to put a flat marble on any pictured word that has a named phoneme or distinctive feature. This is an example of a task in which the field can be expanded to the point that it challenges the child's feature analysis skills. When simplification and substitution phonologic processes overlap, charts can be used to show children how they misperceive and/or misproduce a cluster of phonemes sharing significant features.

For older children and adults, once they show good discrimination of features, or even as they are in the process of improving discrimination, they can use International Phonetic Alphabet (IPA) symbols to represent phonemes, which makes the process more defined. It anchors their perception to a symbol that will not change, no matter what the word is or how variable the spelling patterns the phoneme has. They can efficiently complete an exercise that tests each feature of each phoneme in the language. The charts in Appendix B can be used as reference for exercises for phoneme feature discrimination.

The following list of example exercises represents just a few simple activities that can target specific phonologic perception principles. SLPs are adept at modifying exercises such as these to be specific to the child:

1. Identify the beginning or last phoneme in heard words, pictures, or objects.

2. Identify a numbered phoneme (e.g., third sound in *fish*).
3. Place beginning phoneme at the end of a word/nonword.
4. Count how many phonemes are in a word/nonword.
5. Repeat increasingly-long nonwords.
6. Code with colored chips all the phonemes in words/nonwords with increasingly long and complex CV patterns, using the same color for duplicate phonemes (primary documentation strategy confirming speed and accuracy of phonologic perception).
7. Blend phonemes/syllables of increasing length to identify the word/nonword.
8. Place objects before the child and have the child code phonemes for the word of the object with colored chips, using the same color for duplicate phonemes.
9. Place pictures before the child and have the child code phonemes for the name of the picture with colored chips, using the same color for duplicate phonemes.
10. Say the word and ask the child if the word has a hissing, exploding, lip, teeth, coughing, librarian, or nose sound.
11. Give more than one feature and have the child name a phoneme(s) with those specified feature(s).
12. Name a specified number of words that have a given phoneme in the beginning, middle, or end of words.
13. Name words while in the car that start/end with a (e.g., hissing, exploding, lip, teeth, coughing, librarian, or nose) phoneme.
14. Sort a stack of pictures by a specified phoneme into beginning, ending, and middle word-position piles.
15. Tap and count syllables in words/nonwords.
16. Identify the stressed syllable in words with two or more syllables.
17. Play *Pig Latin*.
18. Play *Clue* by giving some phonemes and their positions and having the child guess the word.
19. Blend phonemes and syllables to identify the word/nonword they represent.
20. Say words/nonwords with a missing syllable/phoneme and have the child guess the word/nonword.
21. Specify one or more phoneme-color matches, for example, red is always [p] and [i] is always white (depending on memory skills), and have the child code spoken words/nonwords using colored chips.
22. Say a sentence(s) slowly and have the child clap every time a specified phoneme is heard.
23. Have the child say which articulator(s) participated in a specified phoneme.
24. Have the child say a descriptor for the beginning phoneme of each word in a sentence (e.g., hissing, coughing, exploding, lip, teeth, librarian, or nose).
25. Reverse roles and have the child give directions to the clinician and see if the clinician follows directions correctly, with occasional intentional errors.
26. Ask:
    - Did your nose tingle when you made that sound?
    - How many sounds make the nose tingle?
    - Which one of these sounds pops or explodes?
    - What is the "twin" of [tʃ]?
    - Which of these sounds hisses?
    - Was your tongue forward or back for [ŋ]?
    - What did your teeth touch?
    - Tell me five features of [f].

APPENDIX D

Phoneme Deletion Exercises

These lists use spelling patterns that are more predictable for sound-letter correspondence but still are not phonetic symbols, allowing adults who are not familiar with the International Phonetic Alphabet (IPA) to pronounce the nonwords. It does not really matter how the nonwords are pronounced as long as any repronunciations match what was said the first time. The child repeats the word after the adult says it and then is asked to say the nonword by not saying the specified phoneme (not letter). The exercise can also be used for color coding sequences of phonemes, first spelling the complete nonword with color coding first and then removing the object representing the named phoneme.

Table D–1. Nonwords With Deletion

	Trial 1	Trial 2
whipsel without p		
lickmik without m		
neyego without g		
walkle without w		
sipkong without ng		
wice without s		
lomp without p		
linz without l		
keyesko without o		
pompoe without m		
wiggie without w		
tempat without m		
zammer without er		
nokeeboo without oo		
laprung without ng		
kissle without le		
wilk without l		
simperso without m		
fangber without ng		
walksoogle without k		
ippersick without ck		
noberdo without er		
wangle without w		
shimzwer without sh		
mannerbee without ee		
nashbok without sh		
hissmat without m		
mocksomo without k		
ikkle without le		
flomp without l		

Table D–2. Nonwords for Coding

sinderp	whipip
dikerdoip	kingkle
rupring	pupering
zingerpop	lieming
feemong	tesering
wakesake	ipsermine
wylingle	apesing
sopebub	akermaze
boopsume	gezmerdope

Common Discrimination Errors

If a child perceives and produces [θ/s], [ð/z], [f/θ], and [v/ð] (a fairly common previous or current articulation error pattern that affects spelling) and has studied the feature differences of all six phonemes but is still making errors, the following chart may be helpful. The clinician pronounces each syllable while the child looks at three pictures of mouths showing interdentalization, labiodentalization, and sibilancy (e.g., Lindamood Phoneme Sequencing [LiPS] pictures or diagrams in Appendix J). The child puts flat marbles on the right side of the picture for the voiceless cognate and on the left for the voiced cognate. The clinician writes responses as correct or incorrect on the chart. This allows tracking of improvement as the exercise is repeated, and it serves as a motivational tool.

Table D–3. Discrimination of [s, z, θ, ð, f, v]

soʊ	θi	voʊ	feɪ
ðaɪ	fa	zoʊ	sʌ
zɛ	θu	ðaɪ	fɛ
vi	sæ	zɝ	θu
ðʊ	fi	voʊ	su
zi	θɝ	ðæ	fɛ
vaɪ	si	zoe	θeɪ
ðʊ	fɛ	vɝ	su
zeɪ	θaʊ	ði	fæ
vu	saɪ	zaʊ	θʌ
ðoʊ	feɪ	vaʊ	seɪ

Figure D–1. Discrimination of [ɪ], [ɛ], and [æ].

Table D–4. Discrimination of [ɪ], [ɛ], and [æ]

igloo	egg	apple
it	led	ant
hid	bell	damp
stick	ten	plan
bill	sell	gas
swim	dead	stamp
fit	bless	tap

Table D–5. Nonword Discrimination of /ɪ/, /ɛ/, and /æ/

mip	mep	map
bip	bep	bap
dit	dep	dap
fip	fep	fap
gif	gep	gap
kiz	kez	kaz
lib	leb	lab
mim	mem	mam
nif	nef	naf
piv	pev	pav
ris	res	ras
tib	teb	tab
vip	vep	vap
zim	zem	zam
biv	bev	bav
miv	mev	mav

Thirty front-vowel words from the chart that follows this form can be said randomly for the child to be asked what the vowel is. If the child answers correctly, a point is earned that is recorded on the form. This exercise is repeated five times with the intention that scores will improve.

Table D–6. Front Vowel Discrimination for [i], [ɪ], [ɛ], [eɪ], and [æ]

	Trial 1	Trial 2	Trial 3	Trial 4	Trial 5
1					
2					
3					
4					
5					
6					
7					
8					
9					
10					
11					
12					
13					
14					
15					
16					
17					
18					
19					
20					
21					
22					
23					
24					
25					
26					
27					
28					
29					
30					
Total	/30	/30	/30	/30	/30

Table D–7. Discrimination of [i], [ɪ], [ɛ], [eɪ], and [æ] in Words

[i] eat	[ɪ] it	[eɪ] vacation	[ɛ] egg	[æ] at
Pete	pin	pain	pen	pan
/p/	pit	—	pet	pat
bean	bin	—	Ben	ban
seen	sin	sane	—	—
wheat	wit	wait	wet	—
wean	win	Wayne	when	—
neat	knit	Nate	net	gnat
—	Jim	—	gem	jam
peek	pick	—	peck	pack
keen	kin	cane	Ken	can
keep	pip	—	pep	cap
beat	bit	bait	bet	bat

Discrimination of Glides

The following words can be presented in random order to distinguish /w/, /r/, and /l/ by pointing to a designated icon for each glide phoneme or coding the words using a specific color for each glide and random colors for other phonemes.

Table D–8. Discrimination of Glide Contrast Words

white	right	light
wing	ring	ling
where	rare	lair
wink	rink	link
walk	rock	lock
weed	reed	lead
whip	rip	lip
wade	raid	laid
woo	rue	Lou

Discrimination of Nasals

Sometimes color coding a list of words that targets omission of nasality will clarify the features. Each of the three nasal phonemes can be designated by a specific color as other phonemes are spelled with random colors.

Table D–9. Discrimination of [m], [n], and [ŋ]

stamp	smoke	sneak	single
single	skunk	sandwich	wrinkle
lunch	spent	smack	swamp
snack	sneeze	pants	whimper

APPENDIX E

NAR Spelling Forms

These sheets can be enlarged for children who need more writing space, but the charts are usually adequate even for kindergarteners when they are printed on 9 × 11–inch paper using back-front printing to keep the activity moving and to conserve paper.

Table E–3 can be placed at the front of spelling/reading pages and checked off as phonemes are completed:

Table E–1. Sound-by-Sound Spelling Sheet

Say Word	Say Sound/Write Letters										Say/Rewrite Word
	1	2	3	4	5	6	7	8	9	10	

Table E–2. Syllable-by-Syllable Spelling Sheet

Tap/Check Syllable	Say/Write Syllable	Say/Write Syllable	Say/Write Syllable	Say/Write Syllable	Say/Write Syllable	Say/Rewrite Word

Table E–3. Completion Record for Phoneme Spelling Lists

Date	Plosives	Date	Glides
	[p] as in *pie*		[w] as in *we*
	[b] as in *boy*		[r] as in *run*
	[t] as in *to*		[l] as in *like*
	[d] as in *dog*		[j] as in *you*
	[k] as in *cat*		**Vowels**
	[g] as in *go*		[i] as in *eat*
			[ɪ] as in *it*
	Sibilants		[e] as in *vacation*
	[s] as in *so*		[ɛ] as in *egg*
	[z] as in *zoo*		[æ] as in *at*
	[ʃ] as in *she*		[ɑ] as in *father*
	[ʒ] as in *measure*		[ɔ] as in *or*
	[tʃ] as in *chop*		[o] as in *hobo*
	[dʒ] as in *joy*		[ʊ] as in *book*
			[u] as in *shoe*
	Fricatives		[ɝ] as in *further*
	[f] as in *fun*		[ɚ] as in *further*
	[v] as in *vine*		[ʌ] as in *above*
	[θ] as in *think*		[ə] as in *above*
	[ð] as in *that*		**Diphonemes***
	[h] as in *hat*		[aɪ] as in *ice*
	Nasals		[ɔɪ] as in *boy*
	[m] as in *my*		[aʊ] as in *out*
	[n] as in *no*		[ju] as in *you*
	[ŋ] as in *ring*		[eɪ] as in *vacation*
			[oʊ] as in *hobo*

*Diphoneme: one vowel phoneme gliding to another.

APPENDIX F

Miscue Analysis

Collecting spelling and reading errors for miscue analysis to identify the basis of confusion should be routine, so the clinician should always be recording errors in essays, worksheets, spelling tests, and other read and written material. Examples of actual errors that suggest the origin(s) of the problem are listed below in the Miscue Analysis Chart with their likely underlying vulnerabilities described as follows:

Table F–1. Miscue Analysis Chart

Target Word	Spelled as	Read as	Miscue
built	but		deleted liquid, remembered /u/ was in word
placed	platen		did not hear [s] but heard /t/; added /en/
little	like		sight-guessed similar-looking, familiar word
three	tree		visual confusion of /t/ and /th/ or confusion between plosive and fricative features
suddenly	surprise		sight guessing based on first letter and word length
elephant	elethuf		fricative simplification; deletion of nasal and stop
cloud	coud		glide (liquid) deletion
people	pepole		incorrect visual recall but remembered word had an /o/
stamp	stap		bilabial nasal deletion next to another bilabial
split	sliwpt		metathesis, liquid simplification, addition
swing		sing	deletion of glide
always		away	deletion of liquid and sibilant
surprise		supper	sight-guessed for length and /s/ and /p/
substitute		suburban	abandoned after *sub-*
house		horse	glide [w/r] simplification in diphoneme, sight-guessed for length and /h/, /s/, /e/

The following Miscue Analysis Log can be used to record the probable category involved in specific errors. When children are older and can be presented with some of the spelling patterns such as doubling consonants and adding prefixes and suffixes, they can participate in the analysis of errors and help decide which spelling rule applies and put the error in the appropriate column. If the rule breakdown occurs frequently enough, it may need more focus, recognizing that orthographic spelling rules often require multiple steps of *if-then* and *inclusion-exclusion* reasoning that may need to be presented slowly.

Table F–2. Miscue Analysis Log

Word	Visual Recall	Spelling Rule	Phonologic	Neglect

Contextual reading errors often reflect the pressure created by the need to pronounce words fluidly, which can result in guessing at words. At least in spelling, children can usually stall for a few seconds to consider options. Children who are prone to impulsivity have a double disadvantage in oral reading. They often require cues to focus, as well as slow rate to pronounce words accurately. In an effort to develop self-monitoring, a baseline should be established for how many errors were made in a passage by making a mark above each incorrectly read or omitted word. The passage can be read many times in an effort to reduce the number of errors and allow reading without interruption. The practice effect that improves fluency is not a problem. It does not negate real gains; it actually facilitates the melody, intonation, and stress pattern of normal reading. Children with dyslexia typically do not accurately perceive the melody and stress pattern onto which they must place the phoneme patterns. Some children, having been instructed that they need to read with feeling, produce exaggerated tonal fluctuation that, nonetheless, shows their lack of familiarity with a normal intonation and stress patterns. Children with dyslexia need time to consider the coding options and pronounce words with correct intonation and stress. Too much focus on speed, couched in the need for fluency, can be self-defeating and actually lead to more nonfluent reading and coding errors. Sometimes syllable pacing, which automatically reduces rate, also reduces the impulsivity so that syllables and words are not skipped.

These are but a few of countless activities that can be used to increase awareness, clarify, and even correct misperceptions about mouth structure, movement, and air emission patterns in relation to the acoustic message that must eventually be coded with letters.

APPENDIX G

Sample Charts, Sentences, and Stories From NAR Lists

After spelling words sound by sound or syllable by syllable, it is important for the child to have a sense of how frequently different letter patterns for each phoneme are used. Sorting the words according to each pattern helps verify that phonemes can be spelled in different ways and which patterns occur most often, but phoneme frequency of occurrence in the Neurolinguistic Approach to Reading (NAR) lists is not an absolute value. Distributions will vary somewhat from grade level to grade level, and words from other sources may show different distributions, but the most frequently occurring patterns generally hold true across all lists. Children do not calculate how often each option is used, but they sense the importance of the option as they consider the vocabulary item and their familiarity with the word.

The difference between spelling words from workbooks and words the child uses in everyday life is that spelling words found in spelling workbooks such as the ones used to generate the NAR lists tend to challenge their vocabulary level. They appear to be a mix of reality and expectation. No effort has been made to lessen that expectation challenge, but it is important to have an understanding of an appropriate NAR grade-level set for spelling/reading accuracy based on the child's integration of his or her own vocabulary with that of the lists. That is part of the rationale for NAR exercises using additional association exercises such as sentence generation and short story construction to clarify meaning and solidify storage.

The semantic stretch needed to produce somewhat logical sentences from words within the lists has value in and of itself, plus the sentences provide opportunity to consider fluency in reading as the sentences are collected and read at intervals. Children can be very creative in how they weave the spelling words into sentences and sentences into a storyline. Afterward, they read aloud the story that they have written or dictated. These activities allow an enjoyable break from more sober coding activities.

The child can write or type the words in the Spelling Option Chart and accompanying contextual writing activities, or the adult can type the word as the child tells the adult where each word goes on the Spelling Options Chart and dictates the sentences and stories. Adaptations can be made to accommodate the typing/writing skills of the child in relation to time expenditure, but as much decision making as possible should be made by the child.

After the charts, sentences, and stories are completed, the child is ready to master the spelling/reading errors made in the sound-by-sound spelling chart, which have been noted on the clinician's copy of the list. Each error is analyzed for phonemes that were misrepresented and/or for which an incorrect letter pattern was chosen. If the error is the letter(s) selection, the letter options should be consulted and the correct option discussed. If the error was due to misperception of the phoneme pattern, colored object coding can be used to clarify the phoneme sequence. The list of words in which the misspelled word should have been placed is reviewed to cue associations. The child might make up a sentence with all of the words on a list that share a particular spelling pattern, which often creates an absurd sentence, for instance, *The rabbit scurried into its burrow* or, *His conclusion was confusion* or, *Dear Joe, clean the meat from your ear*!

In the course of spelling analysis, a few patterns that occur across many phonemes or are more consistent can be discussed, for instance, the fact that many words ending in /or/ that are pronounced [ɚ] refer to people, that [ŋ] never begins a word, or that the sound [i] at the end of a word is often spelled /y/ or /ey/, but these patterns should be brought up as they occur, with other examples generated at the time they occur, not as a rule to be memorized. But once the observation has been noted several times, presenting a list of other words showing the same pattern helps solidify the observation, and recall of the pattern eventually stabilizes. At this later stage in spelling activities, homophones can be introduced and discussed to show that words that sound the same have different meanings as well as the fact that words with identical spelling can mean something different and can be pronounced differently. This can be an opportunity to have some fun with puns by making sentences with words that have multiple meanings and/or spellings, for instance, *I'll take two, too, to give one to Joe.* Challenges can be presented, for instance, asking the child to list four [ʃ] words spelled with /ti/, three with /ci/, and five with /sh/. Clinicians should draw on their creativity to find challenging exercises that can be met, are fun, and reinforce storage. Although these activities can embellish reading and spelling work, if introduced too soon or to children who cannot juggle this much information, the activities can be counterproductive.

Ideally, Standard American English (SAE) would be phonetic, and expectations of near-perfect spelling would be reasonable; however, because phonemes are spelled with variable letter patterns that often exceed human visual recall capability, errors are allowable up to a certain level. The degree to which children can realistically meet that level is different for each child and is based on memory, intelligence, and the severity of the dyslexia. Two of the clinician's greatest challenges are to keep work within the zone of proximal development and to keep therapy moving in a child-affirming manner, not insisting on perfection but reaching toward 90%+ accuracy but less if necessary. Eventually, the child's old paradigms shift, and new generalizations take over, but the parameters that determine that shift are many and defy prediction. Some children have relatively good short-term retention, association storage, and retrieval from long-term stored memory. These children tend to gain spelling/reading mastery that can be readily observed. Others will need more practice, but even for these children, progress should be expected. Parents can become discouraged, lamenting that the child spells error words correctly the night before but forgets them the next day. And some children do seem to have amazing resistance to storing patterns correctly, but they do progress. Clinicians develop a sense of progress, usually discerned from tallies and charts, but that sense of progress should be subjected to standardized testing at intervals.

Work should be kept in a folder or, preferably, a notebook because this allows the work to be documented for collaboration with others who will be working on various stages of the process. It also allows children to see how much progress they are making. The sequence of phoneme selection does not matter. Sometimes cues for the next phoneme to be targeted can be discerned from errors on school papers, errors noted while spelling in other projects, errors noted while targeting another phoneme, or a sense of balance between short and long lists. Small acknowledgments of accomplishment include checking off spelled words, highlighting correctly spelled words, providing even small stickers for each mastered word if a pattern has been difficult, and giving much authentic praise. This is a long process that requires considerable stamina and may need more extraordinary rewards for some children.

Older children and adults tend to find intrinsic rewards as they master spelling and reading, but it never hurts to set a goal with a reward, even for an older person.

Different adults in the child's life can take responsibility for helping with reinforcement of skills. Saturation, but not oversaturation, is the objective. Generally, the more reinforcement children receive, the more steadily they make progress. Parents can repeat sound-by-sound spelling using one or a few lists just completed in therapy, or they can do a simple spelling test check to see how many of the words were not retained and practice those. They can have the child write the word five times, saying each phoneme as it is spelled and then go back later to check accuracy again. At some point, a flash card reinforcement system is usually helpful. The objective is to make the words in the lists part of a working lexicon, not a spelling test on Friday for which words may or may not be remembered. In some ways, the standards for correct spelling for children with dyslexia are higher because of their need to compensate for memory deficit and solidify storage. In actuality, it is rare to reach a level of complete or near-complete mastery, but those working with the child see the progress.

Sample Charts, Sentences, and Stories

The following charts and sentences represent children's efforts to place the word-by-word and syllable-by-syllable spelling sheets into a context to help reinforce and store spelling patterns. The children carry out the chart categorization process for spelling patterns by consulting their completed sound-by-sound spelling sheets and saying in which column the word belongs. Next, they generate sentences using words from the lists and then write stories. Most sentences and stories produced by younger children are typed from their dictation, but with almost no cueing for content, except an occasional *wh-* question cue when they are stuck. Reading their own spelling sheets, charts, sentences, and stories provides a much more meaningful context for reading than a book written by someone else. It also gives them exposure to words that are above their reading level but are within their lexicon because they include them in their sentences, which provides them an opportunity to be cued by their own recall to read more advanced words. They always know the "hard" words they have dictated. Their dictation also gives the clinician insight to their composition potential in writing. Most children quickly learn to use table features on the computer if they type the exercises themselves.

For older children with longer lists, the focus is on the chart, with repeated testing of spelling accuracy in a cyclical pattern over time using different strategies that facilitate their recall: groupings; visualization; bizarre associations; pronouncing words as spelled, not said; prefix, suffix, and root analysis; and other strategies that solidify memory. Sentences and stories may not be as useful because the lists are so long.

Sample Charts, Sentences, and Stories From NAR Lists

The following charts, sentences, and stories provide examples of exercises presented with different grade-level lists:

Table G–1. Kindergarten [ɛ] Sample Chart

1. e	2. ai
red	said
get	airplane
help	
there	
yes	

1. Help me; I'm in trouble.
2. There is a red person.
3. The airplane is flying in the air.
4. You said something.
5. Get my ball, please.

One day a red airplane had smoke on its wing because it got shot. The captain said, "Get out of the plane. There's trouble!" The airplane crashed, but the people got out. The mommies and daddies and grandmas and grandpas helped everyone get out. Yes!

Table G–2. First Grade [aɪ] Sample Chart

1. y	2. i	3. i_e	4. igh
try	find	fire	right
by		ice cream	high
cry			light
			night

1. I try to find a toy.
2. My mom doesn't let me buy what I wanted, so I cry.
3. There is fire up high in the skyscraper.
4. I eat my ice cream in the car right now.
5. I see a light in the sky at night.

One night there was light in the sky. I began to cry when I saw the fire in the city. Then the fire truck came and put out the fire. The next day when it was light, we got ice cream. The ice cream store was right by the toy store. My mom told me I could find a toy to buy.

Table G–3. Second-Grade [i] Sample Chart

1. ee	2. e	3. y	4. e_e	5. ie	6. ea	7. ey
cheek	we	lucky	here	piece	clean	chimney
asleep	believe	carry		policeman	ear	
deer				babies	meat	
free				believe	dear	
meet						
wheel						

1. The man is going to clean the chimney.
2. When a mother deer has girl babies, they are called fawns.
3. When it's Christmas, I believe Santa wants me to sleep.
4. A policeman has a police car with wheels.
5. I kissed the dear baby's cheek.
6. The children are free to play outside here.

One night it was Christmas Eve. Santa was flying his deer, but his deer were too tired to fly anymore. Santa fell out of the sled and fell down the chimney. He got his bag of toys and put the doll babies under the Christmas tree. He went back to another house to put gifts there. I hope he comes back here next year to meet me.

Table G–4. Third-Grade [ɑ] Sample Chart

1. a	2. o	3. al	4. ough	5. aw	6. augh	7. u	8. au
calm	shock	walk	brought	straw	daughter	bulb	cause
salt	hottest		ought	law	taught		automobile
also	model			raw			fault
wander	shot			hawk			
hall	copy			draw			
	body			dawn			
	flock						
	forgot						
	spotted						
	plot						

1. My daughter was shocked, so I helped her calm down.
2. That's the hottest pretzel with salt I ever had.
3. The man did not follow the law and shot a person.
4. I bought a model dress, and it was designed with straws on it.
5. I taught my children how to drive an automobile.
6. I have to do a lot of work, also draw a picture for my science test.
7. I brought eggs for Easter.
8. You ought to take care of your body.
9. It was my fault I ran in the hall.

One day there was a bad man who shot a person at a store in the mall. Everybody wandered around each other because they were so scared. A policeman said, "Stay calm," and arrested the bad man. Everyone went back into the mall to go shopping and have a salty pretzel. One daughter got a pet hawk, and a model got a spotted dress. They all got in their automobiles and lived happily ever after. The end.

Table G–5. Fourth-Grade [ɪ] Sample Chart

1. i	2. y	3. ui
flicker	syllable	quilt
dim	mystery	builder
crib	gym	rebuild
distant		
whiskers		
skill		
wilderness		
ridge		
imitate		
mist		
pity		
crisp		
Africa		
children's		
lip		
icicle		
electricity		
glisten		
medicine		
silk		
tin		
shiver		

(Sometimes fewer words are taken from the list for sentences and stories for various reasons.)

1. The room was dim when the electricity went out.
2. The children were in the icicle palace.
3. I was shivering and my lip was bleeding.
4. Lions have whiskers in Africa.
5. I ate Rice Krispies in the gym.

Once there was a person named Maria who went to Africa. She had a seizure at the hospital, and they gave her medicine. Then she went home and watched a movie. She got better and went to the gym and ate Rice Krispies.

Table G–6. Fifth-Grade [s] Sample Chart

1. s	2. se	3. c	4. ss	5. ce	6. sc	7. z	8. ps
censor	dispense	accessory	assent	enforceable	fluorescent	waltz	psalms
consul		adjacent	caress	masterpiece			pseudonym
episode		censor	casserole	persistence			
insolent		et cetera	delicatessen	precipice			
masterpiece		incentive	depress				
persistence		intercept	embassy				
salve		participation	lesson				
sect		precipice	massacre				
sensitive		susceptible					
susceptible		velocity					
unstated							

(Sometimes sentences reveal that the meaning of words discussed while using the spelling sheets will need more semantic support.)

1. She was insolent to the censor who said that her ideas were not enforceable.
2. The sect was an accessory to the massacre.
3. The velocity was at the precipice of destroying the vehicle.
4. She used a pseudonym for her masterpiece.
5. The delicatessen had my favorite casserole.
6. Depress the button to get into the embassy.

Table G–7. Fifth-Grade [ɝ] and [ɚ] Sample Chart

1. or	2. ir	3. ear	4. er	5. ur	6. urr	7. irr	8. ure	9. ar	10. our
parlor	swirl	earlier	beret	turkeys	scurry	squirrel	agriculture	grammar	encourage
coworker	admiral		stern	blur	furrow		feature	vinegar	
worship	giraffes		coworker	burnt				custard	
armor			fern	lurch				scholar	
governor			kernel	furnish				burglar	
director			alert	purse				cedar	
emperor			governor	sturdy				awkward	
author			emergency	purchase				afterward	
creditor			earlier	hurdle					
hickory			berth	burden					
operator			concern	curb					
effort			herb	hurl					
senator			determine	spur					
			proper	nursery					
			grayer	hamburger					
			imperfect						
			concert						
			calmer						
			nursery						
			timber						
			hamburger						
			properly						

Table G–8. Fifth-Grade [ʃ] Sample Chart

1. sh	2. ti	3. s	4. ch	5. ci	6. c	7. ss	8. ssi
shield	cautious	surely	crochet	specialize	ocean	pressure	mission
crush	patience	unsure	machinery	vicious			admission
fashion	stationary			appreciate			profession
mash	instruction			official			missionary
cashier	notion			gracious			
establish	conversation			electrician			
plush	appreciation			glacier			
blush	caution			social			
flesh	stationery						
mush							
astonishment							
childish							

Table G–9. Seventh-Grade [k] Sample Chart

1. k	2. ck	3 . c	4. cc	5. ch	6. qu	7. que
blackmail	fickle	arc	hiccup	archives	lacquer	boutique
frank		calculation	stucco	melancholy	liquor	plaque
khaki		carcass		monarch		technique
kink		contrast		orchid		
		cram		patriarch		
		curio		technology		
		scaly				

APPENDIX H

"Rules" for Double Consonants

- (V = Vowel Letter and C = Consonant Letter)
- If a prefix ends with the same letter that the word begins with, double the letter (*cooperate, reelection, interrelated, misspell, dissatisfied*).
- If a suffix begins with the same letter that the word ends with, double the letter (*openness, skiing, officially, logically, actually*).
- Certain letters do not double: *h, j, k, w, x,* and *y*.
- When adding a suffix beginning with a V to a one-syllable word ending in VC, double the C (*spanned, biggest, flipped, chatty, nagging*).
- When adding a suffix to a two-syllable word ending in VC and accent/stress is on the second syllable, double the C (*occur/occurrence, confer/conferred, refer/referred, refer/referral*) but not when the accent/stress on the second syllable changes to another syllable (*refer/reference, confer/conference, refer/referential*).
- The C is doubled in a one-syllable word ending in V + /s/, /f/, /z/, /l/ (*huff, kiss, hill, bell, miss, fluff, frizz, buzz*).
- To keep the vowel in the first syllable of the word *short*, double the C (*rabbit, mitten, accompany, commitment, wobble, command, attention, immediate*).

Because of the diverse spellings of American English words brought from other countries and at various times, these patterns do not always apply, but they provide older children a sense of why spelling is so difficult.

APPENDIX I

Diagrams

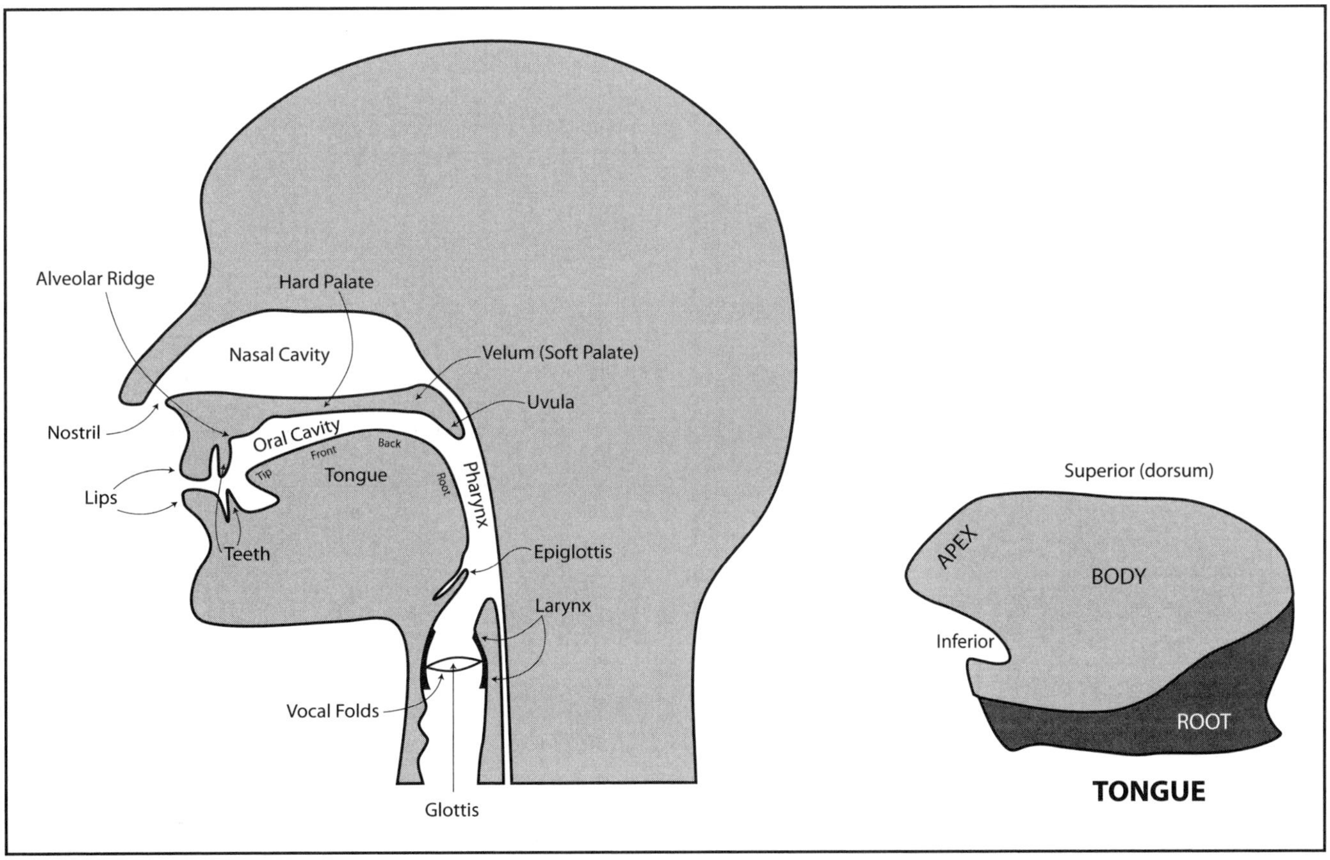

Figure I–1. Anatomy.

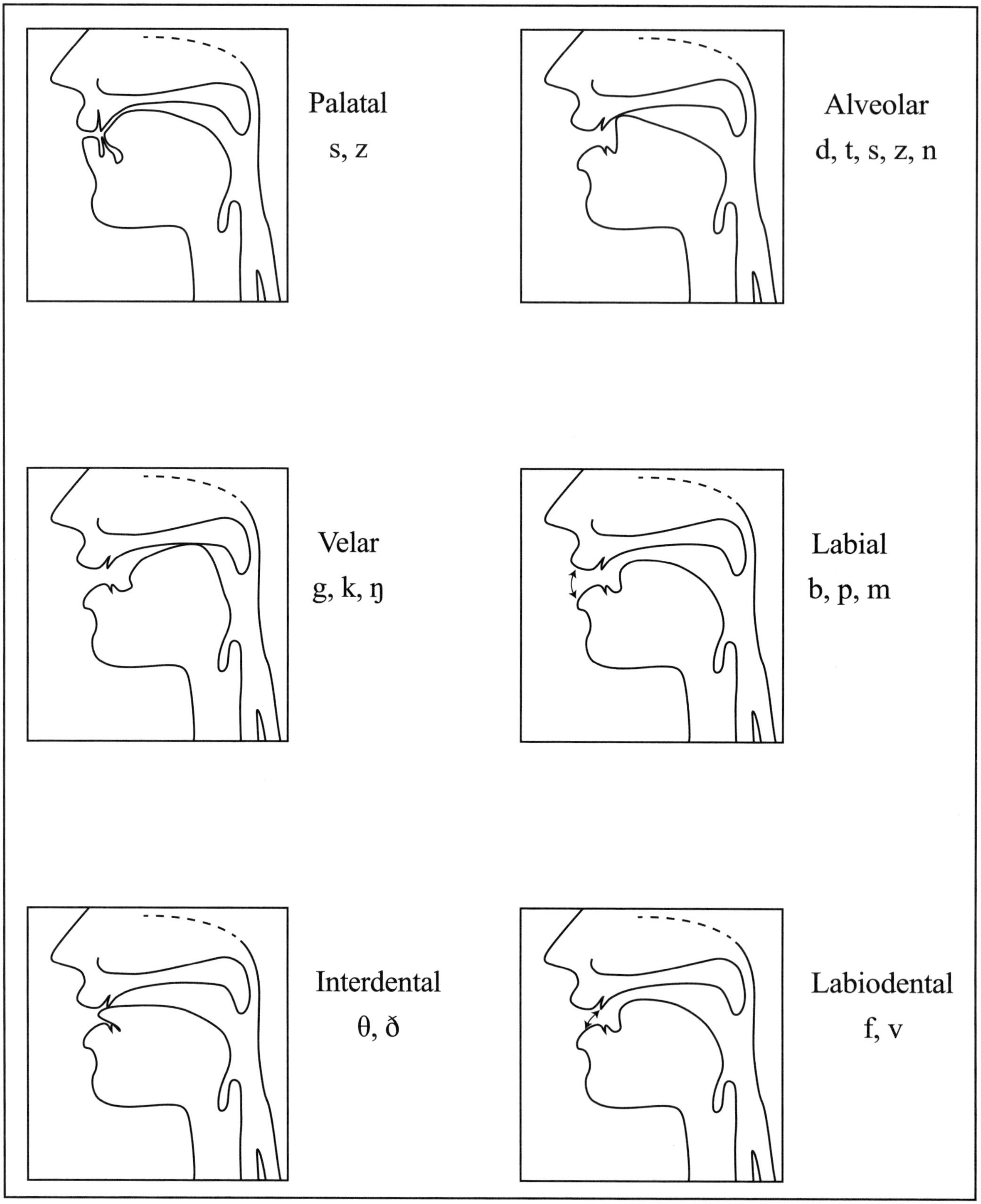

Figure I–2. Articulators.

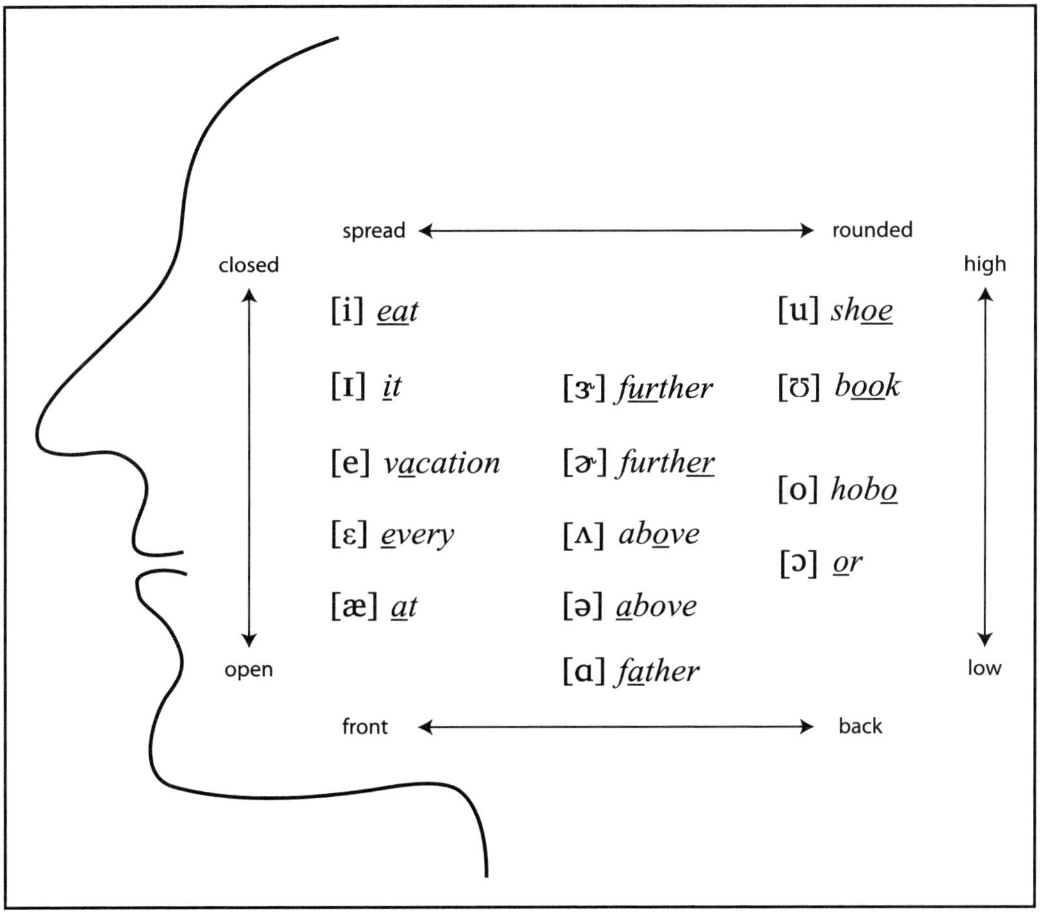

Figure I–3. Vowel schematic.

APPENDIX J

Grade-Level Spelling Sets

Plosives

[p]: p (1)

help
jump
play
stop
up

[b]: b (2)

baby
ball
big
blue

[t]: t (3)

at
fast
it
not
stop
to
toy
want
what

[d]: d, dd (4)

red
good
do
down
daddy
and
did
ride
said

[k]: k, c (5)

car
like
look
can
make
work
come

[g]: g (7)

big
get
go
good
got
green

Sibilants

[s]: s, se (10)	[z]: s (11)
fast stop said see house this yes	is

[dʒ]: j (15)
jump

Fricatives

[f]: f (16)

fast
father
for
fun
funny

[v]: ve (17)

have

[θ]: th (18)

with
something
three

[ð]: th (19)

father
mother
the
there
this

[h]: h (20)

he
help
here
home
house

Nasals

[m]: m	(21)
come	
home	
jump	
make	
may	
me	
mother	
my	

[n]: n, nn	(22)
can	run
now	want
down	
fun	
and	
green	
in	
not	
funny	
on	

[ŋ]: ng	(23)
something	

Glides 

[w]: w, wh (24)

will
we
away
want
what
with
work

[r]: r (25)

car
green
here
red
ride
run

[ɔr] (subcategory of r): or (26)

for

[l]: l, ll, le (27)

look
play
little
ball
will
like

[j]: y (28)

yes

Kindergarten

Vowels 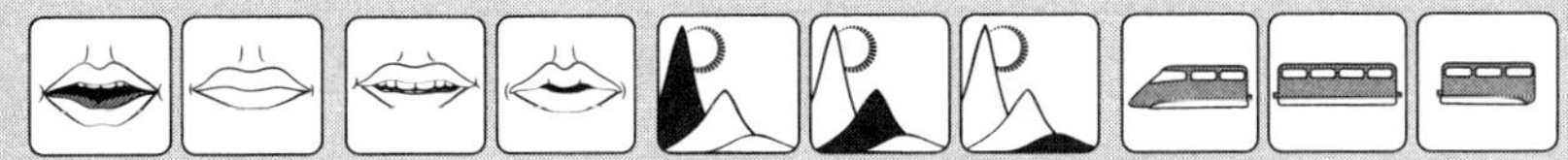

[i]: ee, e, y, e_e (29)

me
here
daddy
green
he
funny
see
be
we

[ɪ]: i (30)

big
did
in
is
it
little
this
will
with

[ɛ]: ai, e (31)

red
get
said
help
there
airplane
yes

[æ]: a (32)

and
at
can
daddy
fast
have

[ɑ] & [ɔ]: a, o (33)

stop
not
want
ball
on

[ʊ]: oo (34)

good
look

continues

[u]: ue, o (35)
do
blue
to

[ɝ] & [ɚ]: er, or (36)
father
mother
work
water

[ʌ] & [ə]: u, o, o_e, a, e (37)	
jump	the
run	
mother	
away	
come	
fun	
something	
up	
funny	
a	

Diphonemes

[eɪ]: ay, a, a_e (38)	[oʊ]: o, o_e (39)
baby away may make play	home go

[aʊ]: ou, ow (40)	[aɪ]: y, i, i_e (41)
down house now	I ride my like

[oɪ]: oy (42)	[ju]: you (43)
toy	you

Sight Words
one
two
are

Grade 1

Plosives

[p]: p, pp (1)	
pen	happy
rope	top
keep	
pack	
puppy	
painted	
deep	
people	
pet	
mop	

[b]: b, bb (2)
bake
bee
tub
behind
rabbit
being
bike
bring
buy
barn

[t]: t, ed, tt (3)	
bite	best
told	
rest	
walked	
take	
looked	
star	
kitten	
gets	
stopped	

[d]: d, dd, ed (4)
dad
rode
find
daddy (K)
hide
called
old
garden
found

[k]: k, c, ck, ch (5)
skip
pink
color
school
neck
kite
back
picnic
cake
basket

[g]: g, gg, gu (7)	
bag	bugs
goat	
dig	
frog	
gate	
egg	
guess	
leg	
pig	
tug	

continues

[ks]: x (8)
box fix fox

Sibilants

[s]: s, se, c, ss (10)	
miss	side
set	cats
hats	asked
circus	
us	
ice	
house (K)	
grass	
let's	
ask	

[z]: s, z, se (11)	
prize	balls
surprise	rose
as	does
please	hens
has	
use	
hose	
pigs	
his	
cars	

[ʃ]: sh (12)
dish
fish
push
she
ship
shoes
shop
wish

[tʃ]: ch, tch (13)
lunch
children
each
catch
much

[dʒ]: j (15)
jet
job
just

Fricatives

[f]: f, ff, gh (16)

far
fly
laugh
further
fit
off
if
calf
life

[v]: ve, v, f (17)

over
gave
of
very
five
give
every
live

[θ]: th (18)

bath
birthday
nothing
path
thank
thing
think
without

[ð]: th (19)

another
other
than
that
their
them
then
these
those

[h]: h, wh (20)

had
hike
hard
hand
hot
who

Nasals 

[m]: m, mb	(21)
man	men
came	am
farm	mom
game	slam
lamb	sometimes
him	time
animal	
met	
plum	
same	

[n]: n, nn, kn	(22)
net	don't
mine	line
train	aunt
know	can't
fine	bone
gone	pony
hen	
penny	
into	
an	

[ŋ]: ng, n	(23)
keeping	sang
living	monkey
seeing	running
painting	playing
uncle	feeling
coming	sitting
long	spring
looking	going
sing	feeding
morning	

Glides

[w]: w, wh (24)	
walk	where
was	
were	
when	
window	
why	
white	
wet	
went	
twin	

[r]: r (25)
ran
hear
read
parade
rug
park
around
near
street
from

[ɔr] (subcategory of r): or, oor, our, ore (26)
or
horse
store
four
story
before
more
door

[l]: l, ll, le (27)	
doll	still
all	black
cold	
apple	
alone	
little	
feel	
sleep	
balloon	
hold	

[j]: y (28)
yard
yes
yellow
yet
year

Vowels

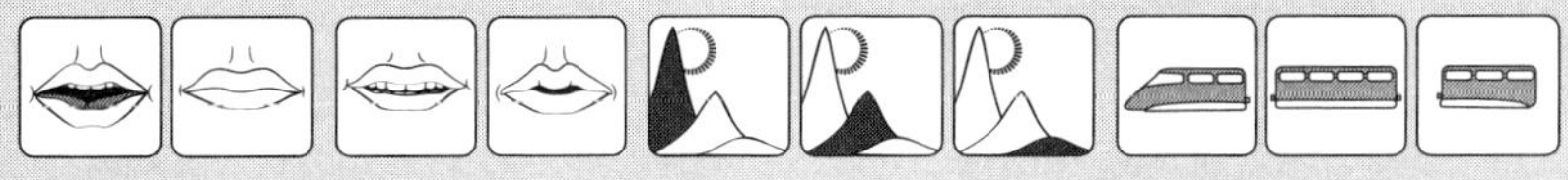

[i]: ee, e, y, e_e, ie, ea, ey (29)

tree
party
hear
me (K)
any
cookie
eat
feet
many
sea
seed
here (K)
honey
feed
pretty
peanut

[ɪ]: i (30)

bit
drip
grin
hill
milk
pick
sit
sits
spin
trick

[ɛ]: ai, e, ea, ie (31)

let
nest
head
friend
again
bed
tell
ready
dress
step

[æ]: a (32)

glad
began
sat
hat
last
cat
bad

[ɑ] & [ɔ]: a, o, al (33)

got
sock
dark
hop
talk
dog
dot
lost
pot
call

[ʊ]: oo, oul, u (34)

would
took
cookie
could
put
book
good

continues

[u]: ue, o, ew, oo, oe (35)
flew
shoe
room
doing
zoo
blue (K)
new
soon
too

[ɝ] & [ɚ]: er, or, ir, ear, ur, urr, irr, ure (36)	
bird	after
faster	her
under	turtle
girl	word
hurry	letter
picture	squirrel
heard	better
never	painter
water	flower
first	

[ʌ] & [ə]: u, o, o_e, a, e (37)		
must	runs	the (K)
cut	but	
around	honey	
sun	truck	
some	hug	
bus	bug	
money	another	
wagon	again	
duck	circus	
nut	woman	

Diphonemes

[eɪ]: ay, a, a_e, ai, ey (38)	
tail	stay
way	name
paint	
made	
baby	
rain	
day	
they	
lake	
rake	

[oʊ]: o, o_e, oa, ow (39)
note
show
so
coat
pole
snow
boat
no
hope
road

[aʊ]: ou, ow (40)
ground
cow
out
about
town
clown
found
our
how
brown

[aɪ]: y, i, i_e, igh (41)
right
fire
try
by
high
ice cream
light
cry
find
night

[oɪ]: oy, oi (42)
noise
boy

[ju]: you (43)
you (K)
your

Sight Words

one (K)
two (K)

Grade 2

Plosives

[p]: p, pp		(1)
suppose	hopped	playpen
open	drop	kept
spell	map	pop
chip	past	paper
puppet	happen	spot
tapped	pail	apartment
cup	peep	

[b]: b, bb		(2)
band	table	
robe	sailboat	
bluebird	rub	
break	subway	
rabbit (1)	able	
library		
maybe		

[t]: t, ed, tt		(3)
until	tape	patted
taken	hit	city
mate	tent	
button	art	
pat	dust	
start	fat	
tie	started	
helped	isn't	

[d]: d, dd, ed		(4)
needed	field	suddenly
sled	puddle	ladder
dirty	filled	dry
need	sad	didn't
carried	middle	
add	drink	
paid	held	
card	planted	

[k]: k, c, ck, ch, cc		(5)
kick	Christmas	careful
second	block	o'clock
candle	key	sink
liked	raccoon	lock
bark	across	
chicken	skate	

[kw] (subcategory of k): qu	(6)
quack	
question	
quick	
quickly	
quiet	

[g]: g, gg, gu		(7)
rag	wag	
angry	guess (1)	
tag	eggs	
biggest	gold	
great	goes	
log		

[ks]: x, xc	(8)
boxes	
foxes	
excite	
six	

Sibilants

[s]: s, se, c, ss, ce (10)	
face	dance
send	dresses
else	glass
bounce	cent
himself	secret
princess	boats
seat	class
base	
nice	
beside	

[z]: s, z, se (11)	
stairs	busy
visit	wasn't
tease	
easy	
because	
glasses	
prize (1)	
houses	
present	
wise	

[ʃ]: sh, ti, s, ch (12)	
flash	rush
sheet	shall
machine	dash
cash	dishes
wash	sheep
finish	station
sure	splash
brush	bush
shout	swish
shiny	wishes

[ʒ]: si (13)
television

[tʃ]: ch, tch (14)	
branch	ranch
teacher	change
chalk	such
kitchen	chipmunk
chair	chin
teach	which
watch	inches
chop	
inch	
reach	

[dʒ]: j, dge, ge, g (15)	
engine	cage
large	jolly
joke	
village	
magic	
bridge	
jug	
edge	
orange	
strange	

Fricatives

[f]: f, ff, gh, ph (16)

fair	soft
wife	elephant
telephone	full
beautiful	beef
wolf	wonderful
office	
safe	
left	
enough	
different	

[v]: ve, v, f (17)

brave	move
drive	of (1)
even	haven't
cover	wave
twelve	saved
waving	ever
leave	leaves
oven	save
I've	cave
glove	

[θ]: th (18)

threw	through
tooth	moth
everything	third
mouth	
both	
teeth	
throw	
fourth	
thought	
north	

[ð]: th (19)

feather
together
brother
grandfather
farther
clothes
grandmother

[h]: h, wh (20)

hair
hay
who (1)
hardly
ahead
hugged
hungry

Nasals

[m]: m, mb, mm (21)	
camp	empty
seem	mommy
hummed	mean
arm	cream
important	remember
climb	farmer
mile	tomorrow
form	I'm
mad	
smile	

[n]: n, nn, kn, gn (22)	
sign	fence
bent	been
seen	corn
wind	knock
knew	neat
tiny	sent
nine	pennies
broken	cannot
dinner	frighten
end	Indian

[ŋ]: ng, n (23)		
song	giving	anything
belong	bank	spelling
making	hoping	talking
hang	hopping	
donkey	jumping	
bang	riding	
cutting	beginning	
king	putting	
eating	building	
evening	ring	

Glides

[w]: w, wh (24)	
whale	wing
week	everywhere
wink	won't
between	we'll
wild	wake
win	
whistle	
always	
wide	
swing	

[r]: r, rr, wr (25)	
roll	wrong
race	fry
radio	roar
write	air
berry	care
pair	tried
part	wear
merry	cross
sorry	
poor	

[ɔr] (subcategory of r): or, oor, our, ore, oar (26)	
short	torn
floor	snowstorm
four (1)	born
board	horn
horses	
popcorn	
shore	
fork	
storm	
stories	

[l]: l, ll, le (27)	
follow	kill
mail	almost
trouble	I'll
bottle	basketball
felt	real
really	flag
clap	only
fold	close
pile	
terrible	

[j]: y (28)
young
yourself

Vowels

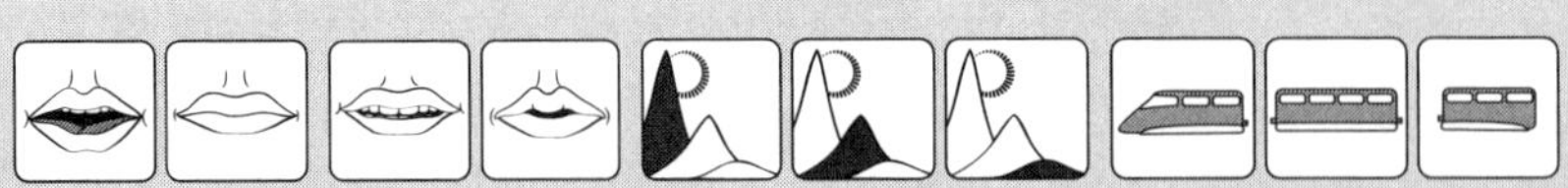

[i]: ee, e, y, e_e, ie, ea, ey, i (29)	
clean	piece
chimney	meat
lucky	policeman
cheek	free
we (K)	babies
here (K)	dear
ear	believe
asleep	meet
deer	wheel
carry	

[ɪ]: i, y, ui (30)	
mill	hid
listen	minute
print	string
bicycle	sick
inside	silly
its	trip
build	swim
ticket	bill
kiss	begin
stick	it's

[ɛ]: ai, e, ea, ie (31)	
friendly	fell
bell	breakfast
ten	well
bread	clever
fairy	
sell	
seven	
heavy	
smell	

[æ]: a (32)	
stand	plant
cap	bat
fan	flat
pan	
track	
family	
crack	
land	
pass	

[ɑ] & [ɔ]: a, o, al, ough, aw, augh (33)		
caught	crawl	strong
moss	pond	lot
rock	fall	cot
clock	fog	tall
wanted	small	paw
trot	bottom	
nod	walk	
wall	bought	

[ʊ]: oo, oul, u (34)	
stood	cook
football	wool
shook	
should	
foot	
wood	
couldn't	
put (1)	

Grade 2 Vowels *continued*

[u]: ue, o, ew, oo, oe, ui, ou (35)	
root	fruit
food	noon
shoe (1)	roof
blew	moon
soup	true
boot	toot
suit	bedroom
do (K)	afternoon
grew	

[ɝ] & [ɚ]: er, or, ir, ear, ur, urr, irr, ure, ar (36)		
picture (1)	bigger	older
tractor	burn	squirrel (1)
number	return	larger
herself	winter	answer
turn	hurt	sister
silver	wonder	summer
world	hunter	driver
birds	early	supper
dollar	hurry (1)	elevator
river	Saturday	
girls	later	
whisper	corner	

[ʌ] & [ə]: u, o, o_e, a, e, ou (37)	
butter	love
trunk	shut
ago	club
luck	unhappy
front	above
loved	idea
country	lion
mud	drum
hunt	afraid
shovel	
stuck	
done	

Diphonemes

[eɪ]: ay, a, a_e, ai, ey, eigh, aigh, ea (38)		
straight	wait	lady
gray	tale	say
neighbor	break	named
chase	rained	place
they (1)	weigh	nail
eight	steak	
baseball	lay	
pay	played	
ate	grade	
today	stayed	

[oʊ]: o, o_e, oa, ow (39)	
row	hoped
railroad	coal
blow	grow
oak	hole
stone	slow
own	cone
most	slowly
raincoat	crow
woke	
low	

[aʊ]: ou, ow (40)	
proud	doghouse
round	owl
drown	loud
sound	
count	
mountain	
flour	
downtown	
mouse	
outside	

[aɪ]: y, i, i_e, igh, ie (41)	
bright	pie
tonight	shine
myself	tired
kind	fight
might	
while	
flashlight	
sky	
pipe	
child	

[oɪ]: oy, oi (42)
join
point
boys
voice

[ju]: you, u_e (43)
mule
useful
you'll

Sight Words

eye
hour
one (K)
two (K)

Plosives

	[p]: p, pp	(1)
company	slept	sweep
crop	carpet	depend
appear	partly	wipe
speak	helper	flap
copper	nap	
lap	happiness	
clip	captain	
pepper	steep	
spirit	lips	
split	group	

	[b]: b, bb	(2)
blade	lumber	February
globe	probably	November
nobody	beat	
bubble	ribbon	
rainbow	bow	
bait	rib	
nibble	bar	
possible	rubbed	
blueberry	tribe	
harbor	unable	

	[t]: t, ed, tt	(3)
mitten	cart	contest
tight	tripped	potato
tube	often	respect
dropped	twenty	beast
faint	attack	route
intend	safety	United States
forgotten	colt	
tune	state	
east	written	
sweet	test	

	[d]: d, dd, ed, t, tt	(4)
trade	dare	Friday
planned	muddy	reddest
decide	butterfly	indeed
fiddle	dried	cardboard
forty	closed	speed
bold	hidden	diamond
dull	batted	toward
drank	lead	badly
matter	cattle	
motor	sudden	

	[k]: k, c, ck, ch, cc	(5)
cabin	tracks	leak
broke	peek	insect
blanket	raccoon (2)	record
welcome	bucket	scarce
correct	spark	case
deck	act	scare
market	strike	
Christmas (2)	nickel	
coconut	poke	
became	cost	

[kw] (subcategory of k): qu	(6)
inquire	
queen	
quite	
squeak	
squeeze	

[g]: g, gg, gu (7)	
bargain	ugly
forget	single
beg	eagle
gift	sugar
struggle	begun
dragon	greedy
guard	dug
signal	guide
baggage	finger
grab	

[ks]: x, xc (8)	
ax	expect
extra	except
excitement	excuse
exclaim	
explain	
sixteen	
exciting	
exercise	
sixty	
mix	

[gz]: x (9)
exactly
example

Sibilants

[s]: s, se, c, ss, ce, sc (10)		
boss	scene	sail
blast	desk	classes
saddest	skin	peace
press	lesson	sale
fierce	certain	slice
mice	sense	stare
scent	rice	closest
toss	blossom	sir
grease	closely	less
chance	space	fasten

[z]: s, z, se, zz, ze (11)		
freeze	puzzle	buses
choose	amaze	says
cousin	flags	doesn't
cheese	legs	size
cozy	gaze	he's
season	wisdom	hasn't
Tuesday	closet	
breeze	lazy	
buzz	chose	
seize	president	

[ʃ]: sh, ti, s, ch, ci, c (12)		
shake	direction	shell
selfish	sharp	shy
share	fisherman	crash
sure (2)	ocean	she's
shape	brushes	
vacation	nation	
delicious	special	
foolish	fresh	
machine (2)	astonish	
shade	shed	

[ʒ]: si, s (13)
treasure
usually
television (2)

[tʃ]: ch, tch, t (14)		
chew	stretch	recheck
speech	butcher	chocolate
bunch	match	search
pitcher	chain	chuckle
patch	coach	benches
fortune	mischief	march/March
hatch	scratch	
beach	cherry	
stitch	bench	
chick	check	

[dʒ]: j, dge, ge, g (15)	
giant	imagine
package	June
judge	charge
average	age
jelly	garage
jar	largest
page	arrange
gentle	stage
July	manage
message	cages

Fricatives

	[f]: f, ff, gh, ph	(16)
buffalo	awful	offer
scarf	fare	fur
stiff	unfair	difference
fifty	telephone (2)	chief
fear	fifteen	cliff
enough (2)	cuff	comfortable
swift	foam	forest
frame	softly	giraffe
traffic	fed	
coffee	loft	

	[v]: ve, v, f	(17)
vegetable	elves	seventy
hive	uncover	everybody
carve	beaver	lively
several	halves	forgive
dive	invite	discover
calves	stove	
of (1)	visitor	
heaven	drove	
lovely	alive	
travel	pave	

	[θ]: th	(18)
breath	thick	
cloth	thief	
earth	thin	
faithful	things	
fifth	thread	
forth	throat	
healthy	truth	
month		
monthly		
south		

	[ð]: th	(19)
bother	they're	
clothing	though	
either	weather	
gather		
neither		
rather		
smooth		
that's		
themselves		
there's		

	[h]: h, wh	(20)
heat	half	
history	reheat	
perhaps	who's	
whose	somehow	
behave	husband	
holiday	who (1)	
beehive	hundred	
hare	hate	
whole	here's	
hoof		

Nasals 

[m]: m, mb, mm, mn (21)		
metal	harm	mild
women	army	common
team	autumn	sometime
hammer	steam	dime
moment	simple	stem
smart	dumb	tremble
mark	mount	somebody
comb	tumble	moved
famous	remain	aim
flame	moonlight	America

[n]: n, nn, kn, gn (22)		
inn	announce	known
since	handle	spend
bean	bunny	American
knife	cotton	
fancy	earn	
tunnel	lawn	
bend	meant	
snake	gnaw	
news	content	
ninety	iron	

[ŋ]: ng, n (23)		
during	sledding	sting
among	leaving	stopping
walking	rang	baking
hung	reading	swimming
choosing	saving	taking
digging	scrubbing	wrapping
getting	skating	blessing
having	smiling	waving
helping	sliding	interesting
tank	changing	sank

Glides 

[w]: w, wh (24)		
west	war	what's
weep	sidewalk	weren't
reward	sandwich	we've
awake	weak	Wednesday
wheels	highway	we're
wire	somewhere	
waste	wider	
forward	switch	
twice	won	
wheat	whip	

[r]: r, rr, wr (25)		
repair	wrote	steer
spray	narrow	pear
grown	prepare	arrive
mirror	nearly	tear
beard	carrot	spare
stream	drawer	grate
wreck	rack	aren't
darkness	rear	
warn	heart	
borrow	scream	

[ɔr] (subcategory of r): or, oor, our, ore, oar (26)	
airport	score
fourteen	course
order	sort
porch	tore
pour	porches
force	worn
outdoors	shortly
wore	
organ	
board (2)	

[l]: l, ll, le, il (27)		
allow	mainly	pupil
belt	leaf	clear
lonely	barrel	cell
finally	lie	laid
sparkle	bundle	pencil
silent	heel	
list	meal	
fellow	pilot	
startle	plainly	
glance	safely	

[j]: y (28)
beyond

Vowels

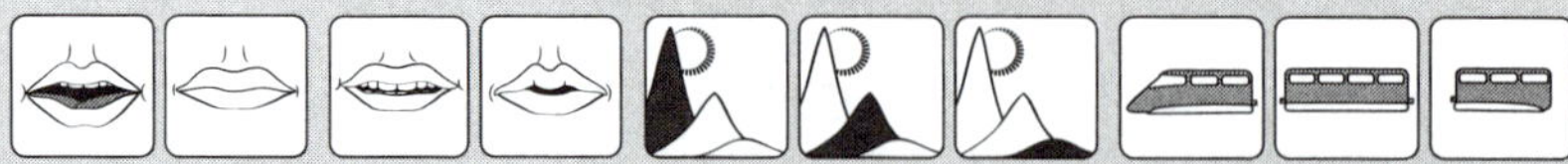

[i]: ee, e, y, e_e, ie, ea, ey, i, ei (29)		
valley	geese	receive
dream	sleepy	reason
police	event	cheer
pea	steal	feast
below	believe (2)	lean
needle	curious	bead
anybody	journey	
piano	repeat	
treat	easily	
pretend	complete	

[ɪ]: i, y, ui (30)		
brick	notice	lived
rid	pin	mistake
tip	impossible	till
lift	refill	
built	rich	
dip	strip	
interest	ill	
sick	pitch	
bicycle (2)	slip	
willing	spill	

[ɛ]: ai, e, ea, ie, a (31)		
led	melt	rented
plenty	dead	parent
unless	itself	shelf
pleasant	downstairs	crept
friendly (2)	carol	careless
again (1)	eleven	
spread	spent	
chest	instead	
already	necklace	
canary	bless	

[æ]: a (32)		
ant	rapidly	understand
apples	rattle	wrap
damp	sand	
drag	satisfy	
gas	slap	
grand	snap	
lastly	stamp	
palace	swam	
plan	tap	
practice	trap	

[ɑ] & [ɔ]: a, o, al, ough, aw, augh, u, au (33)		
calm	taught	body
daughter	raw	ought
shock	bulb	flock
hottest	also	wander
straw	cause	forgot
salt	hawk	spotted
walk	copy	plot
model	automobile	fault
law	brought	hall
shot	draw	dawn

[ʊ]: oo, oul, u (34)
hook
cookies
goodness
bull
understood
brook
wouldn't

Grade 3 Vowels *continued*

[u]: ue, o, ew, oo, oe, ui, ou, u_e, u (35)	
cool	tool
rule	costume
kangaroo	shoot
pool	soup (2)
stupid	loose
prove	lose
fool	spoon
true (2)	
juice	
goose	
crew	
hoop	
canoe	
drew	
rooster	

[ɝ] & [ɚ]: er, or, ir, ear, ur, urr, irr, ure, ar, our (36)			
squirrel (1)	courage	certainly	keeper
thirty	spider	mayor	worry
redder	curtain	master	chatter
glitter	worth	jerk	passenger
worm	herd	honor	tower
closer	furry	adventure	robber
skirt	thirsty	conductor	stir
purple	desert	enter	sadder
cellar	curl	learn	cuter
church	sailor	temper	danger
calendar	shirt	scatter	peddler
circle	deserve	serve	sharper
person	Thursday	ruler	eager
worse	doctor	center	tender
coward	dirt	hotter	batter

[ʌ] & [ə]: u, o, o_e, a, e, ou, oo (37)		
serious	sunshine	study
upstairs	trust	upon
puff	become	Sunday
touch	rubber	camera
cannon	son	apron
bunk	stump	disappear
couple	unlucky	aboard
none	drums	address
struck	gun	against
blood	lump	alarm
stuff	banana	apart
colors	hello (1)	

Diphonemes

[eɪ] & [e]: ay, a, a_e, ai, ey, eigh, aigh, ea (38)

sprain	they (1)	break (2)
tray	April	plate
sleigh	eighty	fail
paste	plane	Monday
pale	straight (2)	May
main	plain	grain
raise	weight	stray
clay	gay	repaint
taste	bare	complain
escape	trail	replay
aid	lately	

[oʊ] & [o]: o, o_e, oa, ow (39)

soap	flow
shadow	roast
lower	owe
float	hollow
soak	toast
post	load
moan	pillow
arrow	spoke
bowl	boast
hero	coast
swallow	

[aʊ]: ou, ow (40)

plow	thousand
hound	power
crown	stout
frown	
trousers	
pound	
cloud	
growl	
loudly	
crowd	

[aɪ]: y, i, i_e, igh, ie (41)

replied	I'd
reply	ripe
polite	sigh
pies	blind
pine	dine
lying	alike
sight	widest
die	tire
delight	
divide	

[oɪ]: oy, oi (42)

soil
joy
spoil
oil
boil
enjoy
disappoint

[ju]: you, u_e, ew, ue, eau, u (43)

usual	refuse
continue	cutest
amuse	unusual
jewel	you've
rescue	
unite	
beauty	
you're	

Sight Words

eye (2)
eyes
hourly
island

Plosives

	[p]: p, pp	(1)
slippers	supply	pup
partner	gasp	repayment
accept	limp	sample
poppy	strap	pump
pupil	slippery	appearance
perfect	persuade	postman
cripple	department	slipper
opposite	papa	planet
grip	compare	camper
happened	triple	pit
dampen	happiest	lamp
newspaper	prayer	
temple	separate	

	[b]: b, bb	(2)
pebble	object	cupboard
beetle	crab	hobby
habit	wheelbarrow	bet
mumble	bay	rob
double	cobbler	ruby
grabbed	basin	stable
bamboo	umbrella	members
bacon	grumble	cable
brain	bury	bid
blubber	marble	
celebrate	web	
gobble	bun	
problem	nearby	

	[t]: t, ed, tt	(3)
grateful	tulip	poets'
instant	poet	Antarctica
hesitate	mattress	
tide	pre-tested	
protest	detective	
helped (2)	retreat	
tilt	untie	
crust	statement	
treatment	type	
distant	tar	
stack	teams	
lit	fact	
tighten	thrifty	

	[d]: d, dd, ed, t, tt	(4)
fade	ordinary	matter (3)
medal	paddle	dock
riddle	pride	pedal
powder	dreadful	gradually
declare	braid	medicine
dandy	saddle	
dried (3)	lid	
maid	reader	
undivided	introduce	
due	forty (3)	
waddle	slid	
modern	ladies	
duty	steady	

[k]: k, c, ck, ch, cc (5)		
character	wicked	preschooler
scamper	blacken	cure
kid	spank	prick
climate	cocoa	echo
freckle	colonel	speck
cast	cane	preschool
coral	lick	creak
stomach	jacket	
rocket	anchor	
skunk	cricket	
moccasin	acre	

[kw] (subcategory of k): qu (6)	
aquarium	squeal
earthquake	squint
equal	
quake	
queer	
quickest	
quietest	
quit	
quiver	
squat	
squawk	

[g]: g, gg, gu, gh, gue (7)		
degree	ragged	buggy
disgrace	wiggle	dignity
anger	pilgrim	crag
stagger	guitar	regular
gallop	god	shrug
groceries	target	glare
plug	giggle	Greek
snuggle	league	tiger
program	cargo	wig
elegant	gull	agreement
ghost	hunger	
guest	magnet	

[ks]: x, xc, cc (8)	
explode	Mexican
express	expert
oxen	succeed
expensive	exchange
experiment	experience
ox	accent
accident	
explanation	
except (3)	
explore	
coax	
mailbox	

[gz]: x (9)
examination
examine

Sibilants

[s]: s, se, c, ss, ce, sc (10)

scientist	spike	somersault
messenger	possibly	lonesome
steak	passed	pants
seal	eastern	stripe
sake	ancestor	fist
peaceful	brass	castle
grease (3)	scar	responsible
western	unsafe	least
success	December	sew
suspect	seat belt	soldier
insist	September	
scenery	sill	
stockade	sisters'	
opossum	softness	

[z]: s, z, se, zz, ze, ss (11)

frozen	praise	citizen
accuse	zipper	Japanese
chimpanzee	used	daisy
blaze	blizzard	graze
disease	scissors	
dessert	magazine	
museum	quiz	
wizard	business	
zero	crazy	
possession	prison	
position	realize	
breeze (3)	easiest	
result	dozen	
zone	rise	

[ʃ]: sh, ti, s, ch, ci, c, ssi (12)

precious	permission	ashamed
shift	sure (2)	Russian
magician	action	she'd
patient	ocean (3)	she'll
shown	vanish	Spanish
shipment	attention	fish's
especially	eyelash	British
condition	musician	Irish
collection	ashes	
bushel	impatient	
motion	marsh	
polish	protection	
parachute	squash	

[ʒ]: si, s (13)

measure
Asia

Grade 4 Sibilants *continued*

[tʃ]: ch, tch, t (14)		
peach	nature	pinch
charcoal	patches	natural
stitches	future	screech
hitch	handkerchief	choke
sandwiches	watches'	coaches
chosen	scratches	china
pasture	charm	French
witch	orchard	Chinese
lunches	approach	creature
chart	champion	
catcher	cheerfully	
temperature	touchdown	

[dʒ]: j, dge, ge, g, gg (15)		
gigantic	badge	engineer
engage	dangerous	bandage
lodge	knowledge	suggest
urge	hinge	plunge
jeep	ledge	sausage
legend	hedge	bulge
range	damage	strangeness
carriage	jam	garbage
gypsy	trudge	cabbage
rage	cottage	arrangement
gem	rigid	
dodge	pigeon	

Fricatives

	[f]: f, ff, gh, ph	(16)
laughter	nephew	fuss
comfort	rough	alphabet
fitness	afford	wife's
finely	frontier	calf's
sheriff	difficult	
suffer	folk	
fin	phone	
tough	friendship	
fiery	powerful	
effect	refrigerator	
defend	rifle	
officer	funnier	
sniff	safer	

	[v]: ve, v, f	(17)
navy	pavement	convenient
invent	canvas	savage
private	of (1)	sleeve
servant	everyday	clover
victory	however	movies
delivery	lever	favorite
solve	avenue	valentine
review	nervous	vase
velvet	carnival	starve
knives	gravy	vine
government	preview	favor
improvement	preserve	
harvest	recover	

	[θ]: th	(18)
tablecloth	seventh	beneath
thorn	sixth	thoughtful
throne	thrift	threaten
strength	toothpaste	eighth
thicken	tenth	thrill
bathroom	health	myth
death	thankless	thump
hearth	healthier	depth
thinnest	thickness	twelfth
thankful	thinner	
arithmetic	bathtub	
ninth	thirst	

	[ð]: th, the	(19)
therefore	northern	
scythe	mother's	
whether	they'd	
altogether	they'll	
brother's		
brothers'		
father's		
fathers'		
southern		
breathe		
leather		
although		

	[h]: h, wh	(20)
hotel	hopeless	helpful
happier	whose (3)	who (1)
overhead	hind	high school
hog	hire	hadn't
anyhow	hut	he'd
whom	honk	
unhappily	who's (3)	
heap	harden	
halt	he'll	
preheat	husky	

Nasals

[m]: m, mb, mm, mn (21)		
minor	jumble	homeless
palm	lame	climber
lemonade	committee	tame
swarm	mama	farmer's
immediately	trumpet	gleam
limb	blame	watermelon
mane	crumb	airmail
autumn (3)	motel	
umpire	basement	
enemy	playmate	
skim	thumb	
rumble	rim	
pumpkin	beam	

[n]: n, nn, kn, gn (22)		
knot	carpenter	nineteen
honest	banner	pain
slender	knee	prince
manner	January	reindeer
inning	kindness	sentence
final	gentleman	knelt
antelope	continent	ancient
bonnet	handsome	manners
sneak	brightness	stolen
gnaw (3)	lane	tan
spine	uneasy	gently
knit	blink	unkindness
transfer	nearest	

[ŋ]: ng, n (23)		
following	link	using
tangle	planning	stocking
sprang	prepaying	jingle
length	rebuilding	swung
gang	kingdom	writing
ink	rewriting	Thanksgiving
setting	dangle	wring
sung	shoving	according
language	jungle	
lightning	singer	

Glides

[w]: w, wh (24)

swept	twist
weed	wharf
wound	anyway
whine	somewhat
swan	weave
anywhere	driveway
beware	wept
swamp	swell
worried	wigwam
weaken	wives'
whenever	twinkle
swoop	weakness
meanwhile	
sweater	

[r]: r, rr, wr (25)

greet	minstrel	fiery
roam	terror	square
wrinkle	weary	area
parrot	trim	mare
snarl	rewrite	pray
quarrel	warrior	trays
streak	direct	darken
pirate	prettiest	arrest
corral	rail	dairy
siren	libraries	driven
horrify	restful	flare
typewriter	wrist	crayon
rare	brake	
married	artist	

[ɔr] (subcategory of r): or, oor, our, ore, oar, ar, aur (26)

hoarse	fortunate	bored
dinosaur	forever	enormous
landlord	floor (2)	forgetfulness
perform	acorn	cord
platform	report	seaport
fort	lord	snort
sport	nor	information
gourd	chore	sore
soar	quart	forehead
quarter	ornament	support
orbit	porter	sword
oar	court	uniform
cork	port	

[l]: l, ll, le, il (27)

bellows	ankle	electric
gallon	pupil (3)	illness
tickle	level	restlessly
steel	lime	tile
bleat	lad	scale
relay	seller	buckle
lone	wail	silence
collect	golden	salute
quill	helpless	howl
glide	settle	female
pillar	peeled	truly
flint	adults	alarm clock
leader	playful	

[j]: y, i (28)

yesterday	canyon
yell	kayak
onion	stallion
coyote	Australia
yawn	Senior
companion	Italian
opinion	Junior
million	

Vowels

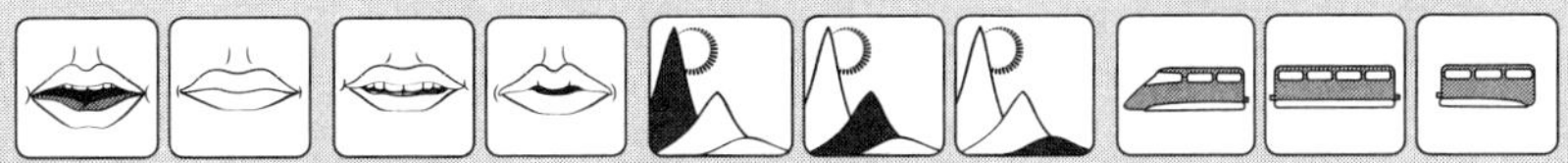

[i]: ee, e, y, e_e, ie, ea, ey, i, ei (29)		
monkeys	brief	beak
spear	eve	peacefulness
freedom	receive (3)	agree
treaty	mysterious	studies
sneeze	teepee	leap
factories	convenience	shriek
diary	peel	deal
ski	armies	deed
relief	tea	Easter
turkey	material	creep
reread	screen	Halloween
uneasily	actually	
funniest	stampede	

[ɪ]: i, y, ui (30)		
quilt	rebuild	Africa
flicker	electricity	
dim	glisten	
syllable	gym	
crib	silk	
distance	tin	
whiskers	shiver	
skill	builder	
wilderness	imitate	
ridge	mystery	
children's	mist	
lip	pity	
icicle	crisp	

[ɛ]: ai, e, ea, ie, a (31)		
editor	unfriendly	cleanse
blend	preparation	represent
weekend	paragraph	restless
dealt	leapt	necessary
den	nonsense	friends
rent	steadily	seventeen
dictionary	lemon	invention
prairie	shelter	self
jealous	emptiness	telescope
general	pretest	debt

[æ]: a (32)		
pant	pad	statue
factory	unpack	tramp
rat	clam	sadness
pattern	slant	pianos
battery	prance	
demand	racket	
acrobat	brag	
grasp	sack	
mast	command	
national	scramble	

[ɑ] & [ɔ]: a, o, al, ough, aw, augh, u, au (33)		
robin	faucet	crocodile
sauce	honesty	shark
unlock	naughty	bob
haul	jaw	daughter's
wobble	astronaut	holidays
claw	pause	August
stalk	shawl	soften
glossary	popular	awkward
fought	wand	October
bulb (3)	drawn	doctor's

[ʊ]: oo, oul, u (34)
woodpecker
painful
wooden
bullet
good-bye
cheerful
should (2)
crooked

continues

[u]: ue, o, ew, oo, oe, ui, ou, u_e, u (35)	
student	stool
moose	mood
scoop	bruise
altitude	cruel
movement	gloomy
broom	suitcase
nuisance	removing
droop	glue
flute	dew
ruin	spool
clue	
remove	
proof	
produce	
rude	
canoe (3)	
improve	
group's	

[ɝ] & [ɚ]: er, or, ir, ear, ur, urr, irr, ure, ar, our (36)			
polar	professor	endurance	churn
term	birch	equator	baker
murmur	squirrel (1)	dreamer	learner
quicker	burst	figure	prisoner
actor	burro	worst	clatter
firm	mixture	speaker	kinder
pearl	particular	earnest	quieter
lantern	inventor	nurses'	writer
furnace	T-shirt	listener	scooter
reporter	herd's	curiosity	settler
janitor	curve	customer	skater
whirl	furniture	hurricane	deliver
encourage	German	banker	stranger
clerk	disturb	turnip	easier
tailor	flavor	hiker	babysitter
burrow	nurse	curly	
capture	alligator	nearer	
collar	bitter	counter	

[ʌ] & [ə]: u, o, o_e, a, e, ou, oo, i (37)		
discovery	stumble	envelope
interrupt	thunder	colony
custom	disgust	instrument
adjust	musket	balance
salad	sum	tomato
punish	buck	encyclopedia
sons'	clump	control
anxious	tuck	mesa
familiar	cub	develop
grunt	shove	flood
hump	current	period
tropical	pioneer	awhile
luckiest	walrus	purpose
mug	couples'	Canadian
ton	happily	rustle
upset	spectacle	

Grade 4

Diphonemes

[eɪ] & [e]: ay, a, a_e, ai, ey, eigh, aigh, ea, ei (38)		
reins	straight (2)	steak (2)
strain	eighteen	
wasteful	sprayed	
dismay	date	
decorate	neighborhood	
obey	placement	
waist	fame	
prepay	contain	
grape	safest	
major	trailer	
stake	payment	
claim	painless	
daily	bracelet	
freight	jail	

[oʊ] & [o]: o, o_e, oa, ow, oe (39)		
owner	toe	scold
doe	slope	shone
poem	goal	oasis
quote	boldness	protect
rodeo	volcano	Eskimo
glow	role	potatoes
pinto	oats	rodeos
oatmeal	poems'	locomotive
stroke	groan	billow
locate	total	
toad	stole	
cove	loan	
shallow	hopeful	
awoke	willow	

[aʊ]: ou, ow (40)	
sour	amount
bound	powerless
brownie	scout
shower	eyebrow
council	surround
crouch	mound
towel	cloudless
vowel	pronounce
trout	playground
scowl	

[aɪ]: y, i, i_e, igh, ie (41)		
design	icy	died
supplies	spied	kindest
grind	entire	fireplace
cycle	stride	fright
higher	flight	child's
spy	fries	slight
wine	dyed	lighten
slyly	remind	midnight
title	admire	
define	brighten	

[oɪ]: oy, oi (42)	
royal	choice
coin	coil
moist	voyage
destroy	disappointment
joint	poison
enjoyment	foil
joyful	moisten
avoid	
enjoys	

[ju]: you, u_e, ew, ue, eau, u (43)	
beauty (3)	argue
unicorn	youth
cube	
unit	
human	
you'd	
useless	
view	
music	

Sight Words

aye	doubt	tongue
Europe	ache	
shepherd	mustache	

Grade 5

Plosives

	[p]: p, pp	(1)
permit	misspell	spade
ripple	spite	depot
maple	mumps	serpent
pardon	plod	multiply
compact	cheap	specialty
oppose	pace	spice
flip	apprentice	palaces
pistol	pep	post meridian
syrup	enterprise	troupe
flop	picket	pal
glimpse	nonstop	cape
hardship	pineapple	poster
hip	chapter	
sparrow	prepaid	

	[b]: b, bb	(2)
bustle	bifocals	
public	oblige	
shrub	cab	
crumble	publish	
tablet	bud	
hobbies	subject	
cranberry		
misbehave		
publisher		
brand		
member		
beets		
humble		
celebration		

	[t]: t, ed, tt	(3)
intent	certainly	rotten
dart	attach	mat
internal	tournament	mister
talent	district	attract
create	construct	perfectly
tack	confident	receipt
tennis	fleet	recent
interested	attempt	recently
alter	gist	restaurant
tourist	treaties	defeat
dropped (3)	consent	motto
credit	instructor	tour
interpret	crest	foster
invitation	cheat	inhabitant
unbutton	butt	

	[d]: d, dd, ed, t, tt	(4)
tinder	defrost	griddle
desired	humid	poverty
pudding	goddess	humidity
midterm	crater	aloud
raid	allowed	patio
gravity	blond	kindergarten
sped	shudder	completed
admitted	capital	permitted
civilized	capitol	midland
gratitude	postdate	
barter	cattle (3)	
radiant	speedometer	
midday	splendid	
Maryland	cities	
midstream	addition	

continues

[k]: k, c, ck, ch, cc, qu (5)		
incline	bouquet	shaken
conquer	brisk	cabinet
chaos	arctic	chorus
karate	suck	practical
scary	tobacco	mask
orchestra	crackle	scald
slick	flake	decree
delicate	cactus	spoken
chemistry	mosquito	risk
checker	mascot	choir
croquet	seek	discuss
parka	kit	scant
scarlet	active	multiplication
kilometer	bookkeeper	
lilac	panic	

[kw] (subcategory of k): qu (6)
quality
squire
equip
unequal
quarantine
quail
liquid
equipment

[g]: g, gg, gu, gh, gue (7)		
gain	congress	granite
ignore	vigorous	rugged
shaggy	nugget	regret
gradual	luggage	vague
neglect	foggy	gap
organize	jog	indignant
bog	regard	jiggle
garment	fig	
spaghetti	progress	
magnify	disguise	

[ks]: x, xc, cc (8)	
explosion	tax
textile	prefix
oxygen	explained
vaccination	excellent
wax	export
perplex	successful
taxi	expression
excited	index
external	exhibition
relax	Texas

[gz]: x (9)
exhaustion
exhaust
co-exist
exhausted
exact
exhibit

Sibilants

[s]: s, se, c, ss, ce, sc (10)

postscript	justice	disagree
brace	science	recipe
muscle	postwar	promise
cinder	assistant	posttest
condense	proceed	recess
tassel	dresser	dismiss
nostril	semicircle	erase
confess	sap	concentrate
advance	mercy	descend
mess	convince	basic
advice	counsel	device
dense	central	hopelessness
descendant	compass	pronunciation
license	chiefs	Massachusetts
essays	recite	criticize
mass	distress	secure

[z]: s, z, se, zz, ze, ss (11)

plaza	fertilizer	wives
chisel	miser	wiser
musical	ease	residential
fertilize	deposit	wares
reasonable	clowns'	transportation
dizzy	companies	wolves
lizard	miserable	heroes
desire	characters	
transport	pansy	
resolution	dazzle	
trapeze	countries	
clumsy	bronze	
scissors (4)	skis	
wisest	invisible	
civilize	author's	
composition	series	

[ʃ]: sh, ti, s, ch, ci, c, ssi, ss (12)

crochet	instruction	plush
mission	official	caution
specialize	ocean (3)	unsure
shield	profession	blush
cautious	cashier	glacier
vicious	gracious	social
surely	electrician	flesh
crush	missionary	mush
patience	pressure	stationary
stationery	notion	astonishment
appreciate	machinery	childish
fashion	establish	
admission	conversation	
mash	appreciation	

[ʒ]: si, s (13)

pleasure
vision
decision
division
occasion

continues

Grade 5 Sibilants *continued*

	[tʃ]: ch, tch, t	(14)
chamber	chant	conscious
perch	unadventurous	
itch	clench	
snatch	latches	
preach	convention	
notch	spinach	
congratulate	clutch	
arch	ditch	
achievement	actual	
recharge	teachers	
sketch	breeches	
century	adventurous	
trench	chum	

	[dʒ]: j, dge, ge, g, gg, dg	(15)
generous	misjudge	courageous
passage	college	badges
rejoice	jewelry	challenge
gentleness	intelligent	illogical
banjo	majesty	imaginary
budge	advantage	intelligence
judgment	energy	agent
lunge	majority	suggestion
energetic	postage	gentlemen
generation	pledge	stingy
jut	pajamas	region
regiment	margin	
mismanagement	project	

Grade 5

Fricatives

	[f]: f, ff, gh, ph	(16)
geography	physical	old-fashioned
refugee	bonfire	café
differ	profit	floe
file	shuffle	failure
orphan	philosopher	cafeteria
trough	confide	
muffle	phonograph	
microphone	certificate	
refuge	atmosphere	
waffle	autograph	
flee	physician	
sulfur	sacrifice	
inform	cough	
triumph	nonfiction	
confidence	beliefs	
fluff	semifinal	

	[v]: ve, v, f	(17)
festival	envelopes	vacant
victim	marvel	Nevada
captive	vanity	vane
traveler	native	nerve
volunteer	grave	various
observe	nonviolent	vary
verse	slave	evil
ourselves	activity	valve
veil	prevent	believe
ravine	previewing	vessel
loaves	provide	vice president
inactive	raven	vast
of (1)	fever	violet
caravan	sever	violin
envious	envy	
creativity	evidence	

	[θ]: th	(18)
thatch	South Carolina	thrilling
theory	athletic	thrive
throb	theater	booth
panther	wealth	thrown
birthmark	thermometer	bimonthly
diphtheria	thermostat	thud
mammoth	thicket	underneath
thimble	broth	warmth
thrust	lengthen	thorough
North Dakota	thoughtfulness	wreath
moth	thrash	

	[ð]: th, the	(19)
seethe		
bathe		
clothe		
thus		
otherwise		
wither		
smother		
soothe		

continues

	[h]: h, wh (20)
hammock	whose (3)
helmet	hustle
whole (3)	Ohio
hiss	height
hockey	prehistoric
holly	who (1)
holy	rehearse
homework	vehicle
who's (3)	whom
huddle	

Nasals

[m]: m, mb, mm, mn (21)		
comment	crime	dam
impolite	numb	elm
grim	scheme	minister
memorial	limit	
immense	slimmer	
bomb	comic	
shimmer	commander	
limestone	male	
solemn	trimming	
perimeter	prompt	
trimmed	column	
cement	nimble	
ammunition	immature	
miracle	terms	

[n]: n, nn, kn, gn (22)		
kneel	unknown	peril
kimono	entrance	darn
mantle	antenna	dandelion
panel	angel	reign
runners	tend	sickness
napkin	canal	sin
connect	nickname	sincere
rinse	lint	snare
foreign	gnat	linen
knob	guinea	undertake
knapsack	nag	centimeter
denied	crane	
mend	goblin	
sunnier	bandit	

[ŋ]: ng, n (23)		
plank	rearranging	drunk
junk	triangle	nonsmoking
flung	cling	accounting
lung	chunk	skiing
boomerang	rung	
blank	rink	
sunk	mining	
clung	junction	
fling	crank	
angle	tinker	

Glides

[w]: w, wh (24)

waterproof	twenty-one
weapon	woven
wedding	wrestle
witness	Wyoming
whisk	Washington
twig	
biweekly	
twirl	
upward	
wedge	
windshield	
wit	
twenty-seven	
wherever	
semisweet	
outwit	
whoop	
swiftness	

[r]: r, rr, wr (25)

rural	darling	remark
shrill	barometer	irregular
increase	terrace	peer
territory	shrink	subtract
librarian	irrigation	berries
macaroni	flair	comrade
rip	terrific	barrette
trial	petroleum	subtraction
correction	original	telegram
tundra	embarrass	art
shipwreck	incorrect	midair
bride	pantry	sorrow
irrigate	horrible	Colorado
errand	career	declaration
rust	prerecorded	ferry
telegraph	presence	carefully
terrify	reed	everybody's
reduce	remarkable	

[ɔr] (subcategory of r): or, oor, our, ore, oar, ar, aur (26)

import	pore
bore	absorb
formal	scorn
fiord	outdoors (3)
coarse	auditorium
gorge	torch
dwarf	semiformal
forty-four	
informal	
normal	
reform	
tortillas	
porcupine	
laboratory	
dinosaur (4)	
glory	
forge	
award	
border	

[l]: l, ll, le, il, el, al, ell (27)

ballet	allowed	racial
liner	visible	parcel
gulp	recall	reel
cradle	stroll	channel
parallel	scalp	sandal
fertile	unluckily	revolutionary
solid	llama	principal
chill	blister	trestle
underline	cereal	calmer
principle	chemical	satellite
millimeter	immobile	easel
pill	pickles	diesel
lack	noble	volley
slash	balcony	gulf
settlement	collie	fossil
salary	pluck	camel
sapling	drill	incomplete
bridle	gravel	farewell
civil	propel	

	[j]: y, i	(28)
yogurt	California	
yarn		
brilliant		
behavior		
lawyer		
genius		
yodel		
yoke		
yonder		
yelp		
New York		

Grade 5

Vowels

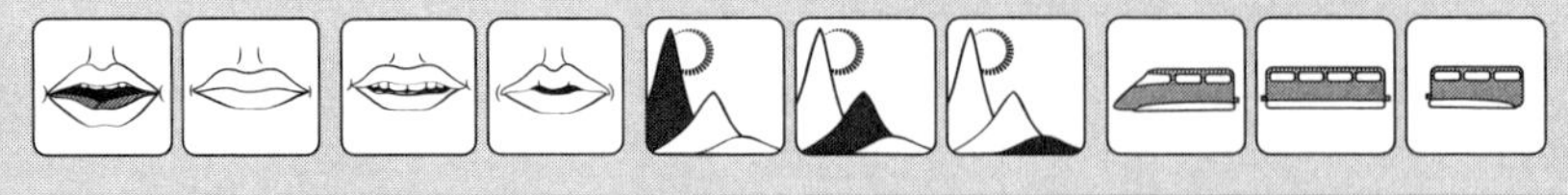

[i]: ee, e, y, e_e, ie, ea, ey, i, ei (29)		
memory	petticoat	seep
shear	breed	glee
niece	reappear	precautionary
melody	county	bacteria
plead	achieve	heal
peak	adobe	barley
bewilder	alley	mighty
ceremony	meek	appeal
concrete	pier	lily
keen	chilly	cheerleader
illegal	merely	academy
pulley	gear	ceiling
decrease	sneer	dreary
pierce	release	peacefully
rally	leash	fury
cease	series	

[ɪ]: i, y, ui (30)	
igloo	tricycle
synonym	rebuild (4)
system	misses
trickle	sheepishly
suspicious	hymn
tinsel	
crystal	
mineral	
liberty	
ridiculous	
bin	
sip	
symmetrical	
symbol	
aluminum	
situation	

[ɛ]: ai, e, ea, ie, a (31)		
sweatshirt	seldom	lair
military	lend	demonstrate
despair	dread	refresh
benefit	relative	sweat
dent	deafen	dentist
semester	desperate	secretary
parakeet	unfriendly (4)	kerosene
reflect	select	refreshment
deaf	election	sombrero
parents	lent	heron
elect	merry-go-round	disinfectant
reflection	homestead	inspect
customary	attendance	
attend	reenter	

[æ]: a, ai (32)		
establish	plaster	valleys
abandon	plastic	captain's
disaster	ram	
dramatic	ransom	
enamel	staff	
fraction	semiannual	
frantic	slab	
task	scrap	
gander	shack	
grant	shaft	
ham	gallon	
Louisiana	staffs	
plaid	standard	
plantation	absent	

[ɑ] & [ɔ]: a, o, al, ough, aw, augh, u, au, oa (33)		
tawny	drawing	sob
frost	harp	stalk (4)
modest	alter	improperly
haunt	sought	launch
tardiness	gauze	cooperate
apricot	honorary	product
install	applesauce	odd
octopus	sprawl	rod
fawn	topic	modernize
sod	par	precaution
bulb (3)	adopt	colonist
false	rotten	helicopter
thaw	populate	prop
stock	monster	loss
broad	population	irresponsible
halter	dishonest	alcohol
audience	olive	naughty (4)

[ʊ]: oo, oul, u (34)
capsule
bully
wouldn't (3)
graceful
book (1)
full-time

[u]: ue, o, ew, oo, oe, ui, ou, u_e, u (35)	
plume	removed
duties	superintendent
stoop	screw
cocoon	schooner
revolution	souvenir
stew	papoose
students	opportunity
improve (4)	dues
haiku	loosen
crude	
shampoo	
harpoon	
canoe (3)	
solution	
sinew	
loom	
include	
bluish	
bloom	
duke	
nuisance (4)	
troop	

[ɝ] & [ɚ]: er, or, ir, ear, ur, urr, irr, ure, ar, our (36)		
turkeys	governor	burden
blur	admiral	effort
squirrel (1)	emergency	determine
beret	purse	curb
parlor	scholar	feature
burnt	director	afterward
stern	giraffes	senator
coworker	earlier	proper
worship	burglar	hurl
grammar	emperor	grayer
lurch	berth	imperfect
swirl	sturdy	furrow
fern	cedar	concert
kernel	author	spur
scurry	creditor	calmer
agriculture	purchase	nursery
vinegar	backward	timber
armor	hickory	hamburger
encourage (4)	concern	properly
alert	hurdle	
furnish	operator	
custard	herb	

continues

Grade 5 Vowels *continued*

[ʌ] & [ə] : u, o, o_e, a, e, ou, oo, i (37)		
molasses	gust	scrub
hover	summarize	gum
stub	bluff	snug
husk	hum	dusk
adapt	spun	monk
strut	gruff	grudge
accompany	instruct	Alaska
dove	nudge	marvelous
blunder	plump	sunniest
rudder	punch	furious
stubborn	interval	aware
lava	independence	
society	reckon	
double-header	affect	
sputter	innocent	
tremendous	ability	
compliment	automatic	
apologize	merchant	
stun	obtain	
gasoline	blood (3)	
scuffle	hull	
instance	Alabama	

Diphonemes

[eɪ] & [e]: ay, a, a_e, ai, ey, eigh, aigh, ea, ei (38)		
wade	strait	saint
stain	vain	gale
eraser	sway	hail
survey	haze	faith
replacement	neighbor	scrape
entertain	rays	bale
ray	hasty	daisies
trace	drain	grace
delay	grayest	entertainment
labor	ape	Maine
operate	dainty	disobey
repay	awaken	break (2)
reclaim	prey	
straighten	detail	
admiration	reins (4)	

[oʊ] & [o]: o, o_e, oa, ow, oe, ou, ough (39)		
froze	colonial	semicolon
bolt	hoe	mold
soul	vote	loaf
dispose	soldiers	homemaker
dominoes	enclose	willow (4)
robot	poncho	co-authors
tone	patrol	sober
tow	dough	code
hogan	growth	boulder
cloak	mole	
elbow	roller	
obedient	sole	
doughnut	polo	
sofa	soda	
decode	rove	

[aʊ]: ou, ow, ough (40)
allowance
warehouse
couch
bough
prowl
drowsy
pounce
fountain
gown
drought
ounce
astound
outfit
account
pouch

[aɪ]: y, i, i_e, igh, ie, ai, ye, y_e (41)		
fright (4)	hygiene	diaries
diameter	migrate	style
stylish	entirely	trisect
miner	sly	skyscraper
bisect	supplied	blindness
outline	iodine	dike
idle	rely	supplying
horizon	aisle	arrival
dynamite	strife	fiber
dial	deny	denying
glider	bind	Iowa
iceberg	dislike	rhyme
typist	trifle	dye
ivory	memorize	
appetite	cyclist	

continues

Grade 5 Diphonemes *continued*

	[oɪ]: oy, oi (42)	
poise	appointment	
loyal	toil	
annoy	unemployment	
moisture	decoy	
loyalty	rejoin	
appoint	employment	
embroider	employed	
oyster		
joyous		
poisonous		
employ		

	[ju]: you, u_e, ew, ue, eau, u, uu (43)	
value	argument	junior
humanity	humor	you've (3)
peculiar	vacuum	circulate
valuable	amusement	education
reputation	cubic	perfume
view (4)	you (1)	community
confuse	university	rectangular
bugle	beautiful (2)	manufacture
fuel	educate	fabulous
you'll (3)	annual	
universe	humorous	

Grade 5

Sight Words	
Connecticut	salmon
headache	
behaviorist	

Plosives

	[p]: p, pp	(1)
paradise	opera	possess
script	impression	preferred
caterpillar	impurity	display
appropriate	independent	propose
aspect	misplace	prow
appliance	opponent	rampart
portable	occupy	reptile
monopolize	imp	campsite
clasp	canopy	response
separately	participant	bumper
comparison	pastime	contemporary
process	peal	shrimp
temporary	per	tropically
periscope	empire	spaniel
slump	perspiration	tarpaulin
evaporate	pint	impure
grope	plea	

	[b]: b, bb	(2)
buffet	dirigible	republic
ambulance	hobble	publication
boom	distribute	beckon
wardrobe	embankment	rabbi
label	hibernate	sensible
rubbish	fable	belly
probable	stubble	stab
baron	globally	album
combat	shabby	capable
vibration	basically	tablespoon
betrayed	bald	variable
turbulent	labels	rebel
suburban	observance	audible
bust	verbally	barb
disability	scribble	
carbon	resemblance	
feeble	absolutely	

	[t]: t, ed, tt	(3)
tyrant	fortress	curtsy
destination	competition	substitute
nitrogen	instinct	superstar
retain	mental	tact
tutor	betrays	taxable
spectator	accomplished	teaspoon
mitt	mutton	complaint
rotary	associate	tentacle
distract	petrify	truant
addressed	poetry	tuna
tense	profited	tusk
notify	raft	attain
constitution	aptitude	typically
stung	satin	abstract
definite	satisfactory	dietary
inattention	sentry	understatement
ultimate	sleet	uncertainty
tunic	baton	carton
entry	stale	

	[d]: d, dd, ed, t, tt	(4)
adore	daydream	presidential
comedian	idol	radios
attitude	edible	additional
cider	forbidden	studios
video	endure	suspend
decimal	federal	introductory
peddle	hideous	litter
boundary	identify	
committed	diet	
latitude	incident	
predictable	introduction	
considerable	tedious	
undeniable	ladle	
aged	cuddle	
contradictory	meddle	
candidate	mid	
dedicate	credited	
residence	clad	
cadet	commend	

[k]: k, c, ck, ch, cc, qu, que (5)		
indicate	occurs	keel
casket	respectable	pecan
accumulate	applicant	archeologist
elk	schedule	recreation
incurable	occasional	lichen
architect	technical	sacred
accommodate	click	democratic
discoloration	mechanic	spectacular
mechanically	antique	skid
calico	acute	tactful
barbecue	characterize	sprinkle
reckless	mock	miscalculation
soccer	comically	conquer (5)
occurrence	echoes	manicure
characteristic	critical	

[kw] (subcategory of k): qu (6)	
acquaintance	squadron
adequate	
banquet	
conquest	
delinquent	
headquarters	
qualified	
qualifies	
quantity	
questionable	
quotable	
quotation	
request	
require	
sequence	

[g]: g, gg, gu, gh, gue (7)		
beggar	disregard	delegate
glossy	goggles	vagabond
agreeable	diagonal	legal
drug	guilty	dignified
magnificent	regulation	biographer
cougar	forgetful	smuggle
migrant	gag	guy
spaghetti (5)	fatigue	disagreeable
obligation	rogue	rig
trigger	spangle	distinguished
monogram	guardian	keg
igloos	dagger	stag
diagram	lag	cigar
catalog	regardless	good-natured
jagged	outrigger	druggist
navigation	investigate	

[ks]: x, xc, cc (8)		
execute	exceedingly	extravagant
expense	accidentally	extreme
excellence	unexpected	extremely
axle	appendix	flax
exceptionally	explanatory	flexible
acceptance	X-ray	axis
exterior	acceptable	nontaxable
exception	inflexible	exclamatory
explosive	expose	expand
suffix	exploration	
excelled	exterminate	
expiration	extinguish	
excess	extract	
extent	approximate	
excelling	extraordinary	
text	extraterrestrial	

[gz]: x (9)
exert
existence

Sibilants

[s]: s, se, c, ss, ce, sc, z (10)		
forcible	absence	cigarette
impress	priceless	assist
courtesy	decent	suspense
slit	sprint	cypress
classify	consist	self-addressed
resistance	ascent	sleek
senseless	percent	necessity
hostess	satisfaction	essay
specimen	cancel	site
stereo	bass	responsibility
lasso	supersonic	substandard
waltz	acid	solo
fascinate	centralize	monstrous
strict	besides	species
loosely	scarcely	
gossip	crisscross	
substance	disastrous	

[z]: s, z, se, zz, ze, ss (11)		
grizzly	opposition	representative
laser	hazel	lens
unpleasant	possession (4)	refers
hesitant	recognize	ooze
bazaar	measles	
patriotism	doze	
resulting	wheeze	
applause	visualize	
cruise	nozzle	
zigzag	personalize	
imprison	pose	
devise	citizenship	
zeros	presents	
advertise	capitalize	
resist	resemble	
irresistible	muzzle	
daze	stereos	

[ʃ]: sh, ti, s, ch, ci, c, ssi, ss, si, xi (12)		
ambition	oceanographer	prediction
relationship	commission	shame
harsh	smash	ratios
mansion	ration	shortstop
financial	pediatrician	contraption
completion	sheer	essential
brochure	detention	application
decoration	antisocial	assurance
lash	procession	emotion
racial	rational	disposition
initial	bishop	artificial
motionless	additionally	friction
accomplish	sash	affection
issue	insurance	contraction
cushion	complexion	
partial	nutritious	
chute	flush	

[ʒ]: si, s, g, x (gʒ) (13)
collage
erosion
leisure
casual
genre
visual
measurement
luxury
provision
incision
confusion
conclusion
revision
occasionally

Grade 6 Sibilants *continued*

[tʃ]: ch, tch, t (14)		
mature	achieves	adventuresome
research	hitchhike	lecture
posture	departure	
venture	chat	
hatchet	latch	
amateur	drench	
signature	enchant	
duchess	ostrich	
charity	premature	
wrench	archery	
unfortunate	starch	
immaturity	fetch	
twitch	wristwatch	
architecture	crunch	

[dʒ]: j, dge, ge, g, gg, dg, d (15)		
fudge	jest	endanger
tangerine	mileage	conjunction
avenge	vegetation	smudge
barge	jay	wages
injure	storage	porridge
biology	register	generosity
rummage	banjos	illegible
graduation	revenge	indigestible
fringe	forgery	religion
pageant	badger	injury
jury	changeable	imaginable
tangible	magically	gesture
jeans	jolt	origin
eligible	siege	suggestion (5)

Fricatives

[f]: f, ff, gh, ph		(16)
pheasant	triumphant	profitable
formation	forfeit	qualifying
sufficient	trough (5)	asphalt
indifferent	photo	confers
hyphen	conflict	misinformation
referred	fir	overconfidently
draft	fate	
affair	photography	
baffle	interface	
orphanage	graphic	
refine	alfalfa	
preface	alphabetical	
sophomore	misinform	
fatal	phantom	
telephoto	definition	
differed	geographer	
offense	nonfat	
flit	fulfill	

[v]: ve, v, f		(17)
govern	evident	devil
prevention	favorable	reservoir
televise	relieve	of (1)
elevation	beloved	reverse
convict	immovable	advancement
voluntary	individual	slavery
conservation	invade	subdivision
convert	ivy	supervise
revive	levy	tavern
devotion	liver	grieve
disadvantage	mischievous	undercover
aviation	naval	varnish
divisible	obvious	vest
attractive	advisable	villain
enviable	oval	virus
environment	overcast	vivid
eventful	positive	vocally
eventual	provides	

[θ]: th	(18)
athlete	wealthy
athletically	worthless
authority	wrath
blacksmith	
cathedral	
enthusiasm	
mathematics	
method	
mythology	
oath	
sympathetic	
sympathy	
underneath (5)	

[ð]: th, the (19)
furthermore
lather
weatherproof
seethe (5)

	[h]: h, wh	(20)
hesitation	whom (4)	
hasten	hornet	
behold	hospitalize	
wholesome	host	
dehumanize	withhold	
handy	inherit	
hangar	who (1)	
hanger		
hem		
herald		
whole (3)		
behalf		
hinder		
inheritance		

Nasals

[m]: m, mb, mm, mn (21)		
amazement	dummy	democracy
impulse	elementary	mermaid
emblem	fumble	demon
summit	primary	commence
harmonize	estimate	permissible
modeled	ceremonies	midsummer
assemble	recommend	preliminary
plumber	dismal	complimentary
remedy	image	ramp
submit	clamp	brim
stammer	communication	remained
cramp	tempt	atomic
condemn	accomplishment	slim
intermediate	manure	immigrant
medium	marshmallow	amen
homonym	criminal	intramural

[n]: n, nn, kn, gn (22)		
ponder	misconnection	maintenance
honorable	economics	kennel
hint	knack	contribute
ignorance	denies	ninety-nine
campaign	cancels	bond
international	combine	panels
connection	resign	permanent
enable	funnel	assign
sonar	vanilla	punishment
menace	conscience	alternate
indent	infant	defense
innocence	continual	astronomy
notable	intentionally	
nestle	senior	
recondition	conduct	
tenant	ascend	

[ŋ]: ng, n (23)		
rank	admitting	obtaining
editing	mink	distinct
interchange	directing	outstanding
portraying	distinguish	paneling
occupying	shingle	permitting
complaining	employing	lanky
kangaroos	slang	preferring
referring	forbidding	profiting
anxiety	canceling	providing
cunning	herring	differing
conferring	imagining	remaining
fang	gangplank	requiring
betraying	introducing	separating
committing	labeling	ailing
completing	advertising	aqualung
continuing	modeling	
crediting		

Glides 

[w]: w, wh (24)	
swollen	worrisome
wasp	waiter
weird	where's
walnut	downward
skyward	ward
twine	dwelling
whimper	swap
wallet	westward
midway	homeward
wondrous	superpower
steward	swat
wheelchair	eastward
warily	inward
swear	
underwater	
railway	
wart	

[r]: r, rr, wr, rh (25)		
fragrant	wren	prearrange
enrage	extracurricular	conference
requires	imprint	reef
barren	overrated	terrain
hereditary	cartoon	clearance
rapid	reliable	realistically
narrator	intrude	rhythm
contract	apparently	difference
portrays	lark	indifference
rhythmically	laurel	quarry
spiral	aircraft	rut
contrary	irrational	rate
shred	misread	interfere
arise	remembrance	submarine
derrick	nonreturnable	fearless
erect	barely	broad jump
marriage	phrase	raspberry

[ɔr] (subcategory of r): or, oor, our, ore, oar, ar, aur (26)		
resort	ford	ore
forbids	foreman	corridor
stork	forlorn	performance
misfortune	dinosaur (4)	porcelain
ashore	gorgeous	pork
immortal	historical	dwarf (5)
mooring	horrid	restore
mortal	corps	retort
sworn	importance	snore
proportion	inventory	accord
dormitory	starboard	chord
core	moral	source
escort	accordion	suborbital
adorable	observatory	
forbid	orally	
territorial	mourn	
overboard	swore	

[l]: l, ll, le, il, el, al, ell (27)		
illustration	mellow	oral
labeled	elaborate	console
pulse	accelerate	plus
enlist	gully	clamber
trolley	selection	ballot
qualify	isle	gallant
leadership	idolize	rebels
privilege	shellac	film
skillful	lassos	helm
politics	apply	canceled
desolate	directly	lull
ally	violence	slush
pelt	flea	supplementary
cultivate	lest	conclude
alas	indelible	deflate
electrically	bleed	almond
relation	mislead	satellite (5)

continues

[j]: y, i (28)
yeast
yolk
yacht
yield
billion
youngster
rebellion
yam

Vowels

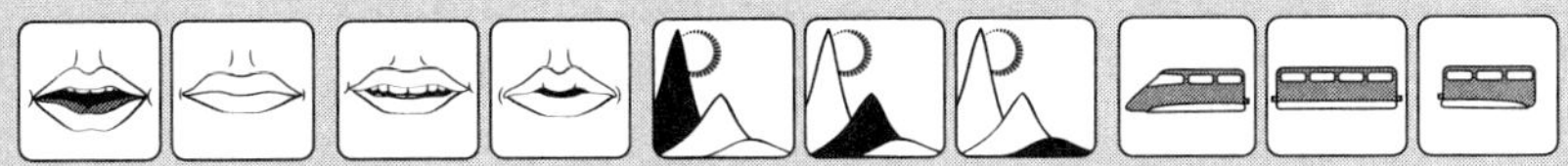

[i]: ee, e, y, e_e, ie, ea, ey, i, ei (29)

beacon	meteor	disagreement
grief	ideally	resent
completely	conceit	galley
geology	fleece	supreme
capacity	allegiance	antifreeze
immediate	belief	reelect
conceal	deceive	priest
patios	seam	superior
alien	feat	eel
heed	beech	interference
crease	rebelling	smear
marine	needy	prefers
referee	achieving	unreasonable
courteous	peacock	security
reap	obedience	

[ɪ]: i, y, ui (30)

midst	idiot	political
kindle	continuous	religious
insult	inflict	shilling
cylinder	mint	sling
missile	intelligible	spit
vicinity	critically	incidentally
description	manuscript	admissible
pitiful	biscuit	
nick	superstition	
armistice	consideration	
civilization	artistically	
physically	typical	
disk	optimism	
assistance	pessimism	
frisky	edition	

[ɛ]: ai, e, ea, ie, a (31)

momentary	fled	peg
readily	fret	pest
genuine	bearable	plentiful
defendant	friendly (2)	professional
respectful	accented	prospect
chemist	heroine	lair (5)
credits	incredible	recession
intense	inspection	correspondent
headlight	cemetery	remedies
duet	intersect	resident
separates	meadow	affectionate
legible	lieutenant	senate
breast	majestic	spectrum
reference	contents	skeleton
attendant	medical	tension
edits	directed	contestant
element	offend	celery

[æ]: a, ai (32)

accurate	handicap	scan
actress	imagines	shatter
aft	impractical	slack
atom	invalid	snack
attraction	italics	span
broadcast	lance	strand
campfire	manager	tackle
chap	mankind	tactless
contact	mechanical	underpass
detract	nationally	unravel
elastic	paneled	van
factually	passerby	
fanciful	practically	
forecast	radish	
gallery	rap	
plaid (5)	sag	
graduate	sarcastically	

continues

[ɑ] & [ɔ]: a, o, al, ough, aw, augh, u, au, oa (33)		
resolve	folly	revolve
solitary	laundry	respond
cockpit	socket	rot
applaud	strawberry	offered
correspond	patriotically	blot
slaughter	saucer	involve
profits	astrology	fought (4)
combination	barber	tropics
lard	improper	dissolve
auction	slot	stalk (4)
bulb (3)	mob	costly
broaden	ecology	aristocracy
logically	nonprofit	geologist
vault	politician	cross-country
models	awe	
auditory	depository	

[ʊ]: oo, oul, u, ou (34)
soot
should've
childhood
hood
detour
could've
overlook
bulletin
bookstore
would've
checkbook

[u]: ue, o, ew, oo, oe, ui, ou, u_e, u, eu (35)		
loop	approval	studio
unsuitable	presume	brood
numeral	scuba	overrule
prune	roommate	suitable
coupon	shoe (1)	routine
influence	caribou	dutiful
pursuing	spruce	fluid
residue	duel	disprove
doom	groom	
neutral	introduces	
who'll	studious	
boost	screw (5)	
fruitless	assume	

[ɝ] & [ɚ]: er, or, ir, ear, ur, urr, irr, ure, ar, our, err (36)

similar	interior	insert	radiator	fender	astronomer
convertible	kerchief	lurk	reversible	aster	snicker
administrator	emerge	vulture	squirrel (1)	garter	itinerary
turf	circular	geranium	vigor	commuter	elder
observation	occurring	spurt	refer	employer	taper
chirp	vapor	monitor	disturbance	attorneys	autobiographer
occur	confirm	commercial	superconductor	meter	terrier
liar	impersonal	pursues	submerge	moderate	interact
sermon	urgent	clamor	durable	literature	
discourage	ignorant	conferred	determination	plural	
odor	directory	curse	assure	overachiever	
operation	personality	navigator	earlier (5)	photographer	
reserve	burglary	prefer	founder	bulldozer	
turban	error	absurd	pliers	primer	
literary	merchandise	rumor	illiterate	flourish	
mortar	surf	impersonate	layer	revolver	
squirt	aviator	resurface	ember	splinter	

[ʌ] & [ə] : u, o, o_e, a, e, ou, oo, i (37)

combustible	butler	bulk	income	accustomed
adjustment	industrious	attendants	couples	apologetically
destruction	camouflage	crutch	adult	
patriot	muskrat	summon	structure	
summary	commits	catsup	dungeon	
preparatory	obstruct	parliament	personal	
compliment (5)	plunder	item	partially	
custody	gutter	melon	legion	
reassure	shutter	industry	ruffle	
pulp	discussion	approve	objection	
slumber	skull	suspension	abundant	
production	numerous	sponge	development	
distrust	slung	mustard	arena	
resulted	uproar	considerate	completes	
avalanche	sprung	flood (4)	miniature	
grub	circumference	secondary	mistrust	
deduct	strung	previous	session	

Diphonemes

[eɪ] & [e]: ay, a, a_e, ai, ey, eigh, aigh, ea, ei (38)			
occupation	estate	maintain	steak (2)
imitation	quaint	interstate	debate
crate	patriotic	obtained	acquaint
complained	slay	drape	naked
bade	sleigh (3)	available	overtake
decay	relate	raisin	intrastate
snail	reins (4)	detain	
complicate	hay fever	fake	
pane	radar	lace	
location	betray	donate	
portrayed	slain	proclaim	
inflate	phase	insane	
frail	prevail	disdain	
straighten (5)	shave	replace	
sensation	surveys	gaily	

[oʊ] & [o]: o, o_e, oa, ow, oe, ou, ough, eau (39)		
promotion	mow	stow
dome	local	halos
approachable	plateau	compose
silos	grown-up	homely
potion	commotion	devote
custodian	lope	
bureau	romantic	
dose	grove	
dough (5)	impose	
promote	scroll	
sow	remote	
strode	association	
solar	coincidence	
foe	poultry	
subtotal	fro	

[aʊ]: ou, ow, ough (40)	
background	devour
bounty	arouse
brow	half-hour
outrage	
doubtful	
mountainous	
fowl	
foundation	
browse	
compound	
household	
drought (5)	
foul	
blouse	

[aɪ]: y, i, i_e, igh, ie, ai, ye, y_e, ui (41)		
hydrogen	vibrate	crisis
itemize	recline	acquire
livestock	frightened	guidance
scientific	cyclone	dice
variety	ideal	confine
inspire	decline	aisle (5)
retire	rival	attire
assignment	defy	prime
nylon	occupies	vice
rye	rhyme (5)	mite
brier	nigh	requirement
perspire	delightful	
nightmare	conspire	
purify	overtighten	

[oɪ]: oy, oi (42)
employee
annoyance
asteroid
disloyalty
hoist
boiler
employs
avoidance
enjoyable
broil

[ju]: you, u_e, ew, ue, eau, u, uu (43)		
refuel	manual	vacuum (5)
utensil	abuse	refusal
computer	youth (4)	communicate
curfew	beautiful (2)	occupied
volume	fuse	monument
hue	document	
interview	misuse	
tribute	musically	
pew	immunize	
union	continues	
dispute	you (K)	

Plosives

[p]: p, pp	(1)	
campus	apologies	cooperative
respiration	drapery	appendicitis
compel	coupe	impossibility
supposed	participate	disproved
collapse	relapse	penetrate
compartment	inscription	monopoly
eruption	lapel	centipede
complicated	pare	duplication
paternal	postponed	persist
nipple	optimistically	pension
depart	parch	

[b]: b, bb	(2)	
lobby	fabric	Republican
hobo	compatible	bribe
asbestos	bask	alibi
ebony	jubilant	
bliss	abdomen	
imbalance	limber	
suburb	tuberculosis	
burial	lullaby	
labored	mobile	
disable	balconies	
obstacle	publicity	

[t]: t, ed, tt	(3)	
trio	clot	temptation
patrolled	reject	sanitarium
etiquette	matinee	admittance
alteration	patent	transcontinental
talon	commitment	hostile
intention	quadruped	turnpike
wrapped	attainable	critic
canteen	rental	testify
entitle	timetable	
vital	dictation	
forgettable	tattoo	

[d]: d, dd, ed, t, tt	(4)	
jetty	Democrat	nude
deputy	brittle	putty
conceited	digest	particle
belittle	discard	periodically
fad	approved	addition (5)
caddie	fundamental	postdated
predict	holdup	laden
omitted	credible	blindfold
academically	kidnap	sardine
petal	premeditated	ballad
crusade	misunderstand	

[k]: k, c, ck, ch, cc, qu, que	(5)	
boutique	curio	monarch
patriarch	fickle	archives
kink	balk	cram
melancholy	hiccup	plaque
carcass	khaki	scaly
arc	technology	calculation
stucco	contrast	technique
orchid	liquor	lacquer
frank	blackmail	

[kw] (subcategory of k): qu	(6)
quest	quotient
ventriloquist	
quadrangle	
equality	
quartet	
quench	
consequence	
quilted	
quantification	

[g]: g, gg, gu, gh, gue (7)		
jogging	centigrade	progressive
grandeur	gloat	gauge
regrettable	dialogue	agony
vulgar	guarantee	smuggle (6)
spaghetti (5)	ingredient	forgivable
anguish	slogan	stenographer
cog	magnanimous	engrave
glint	begrudge	
fragment	plague	
migration	progression	

[ks]: x, xc, cc (8)	
extension	expel
expectant	expired
exceed	accidentally (6)
climax	extinct
approximately	sex
exclamation	
exclude	
excitable	
exhale	
acceptance (6)	

[gz]: x (9)
exile
exaggerate
executive

Sibilants

[s]: s, se, c, ss, ce, sc, z, ps (10)

sane
testimony
census
isolate
psalm
misspelled
arsenal
slime
circumstances
psychology
descent
genesis
absence (6)
loosely (6)
massive
adolescent
colossal
psychic
suite
reception
access
waltz (6)
serum
assassinate
specifically
scenic
statesman
sue
superlative
ace
transom
spire
crescent

[z]: s, z, se, zz, ze, ss (11)

proposition
razor
vise
despise
buzzard
scissors (4)
pasteurize
jazz
clause
horizontal
influenza
wheeze (6)
misery
desirable
sterilize
crimson
flimsy
representation
chasm
mosaic
supervisor
baptize
displease

[ʃ]: sh, ti, s, ch, ci, c, ssi, ss, si, xi (12)

insure
freshman
regression
ambitious
issue (6)
marshal
initiate
carnation
bashful
complication
banish
complexion (6)
ferocious
chef
infection
ambush
initiation
emotionally
shrivel
mansion (6)
chauffeur
mesh
oceanographer (6)
passion
punctuation
efficient
shrine
chivalry
suspicion
tradition

[ʒ]: si, s, g, x (gʒ) (13)

luxury (6)
exposure
corsage
precision
espionage
version
immeasurable
mirage
excursion

[tʃ]: ch, tch, t (14)

satchel
perpetual
chairman
maturity
sculpture
chapel
bachelor
parchment
influential
cultural
contagious
treacherous
digestion
batch
cherish
roach
picturesque
chess
adjoin

[dʒ]: j, dge, ge, g, gg, dg, d (15)

prejudice
foliage
gadget
ajar
juvenile
graduation (6)
objective
suggest (4)
acknowledgment
fudge (6)
geological
longitude
frigid
refrigeration
jeer
tragic
prestige
logical
jilt
manageable
midget
indigestion
genealogy
injection

Fricatives

[f]: f, ff, gh, ph (16)

hemisphere	sphere	suffered
preferable	infect	acrophobia
factor	pamphlet	flask
trough (5)	belfry	efficiency
sift	phobia	
emphasis	freak	
tariff	rafter	
emphasize	scoff	
frankfurter	phenomenon	
symphony	feud	
graph	stereophonic	

[v]: ve, v, f (17)

of (1)	levee	violent
reveal	novel	involving
vitamin	investigation	civics
cavern	dissolved	vulcanize
starvation	province	
intensive	advancing	
vomit	salvage	
conviction	divine	
oversight	survival	
vent	crevice	
lavender	via	

[θ]: th (18)

sympathize
thresh
filth
mirth
sheath
thigh
philanthropist
enthusiastic

[ð]: th, the (19)

tether
smother (5)
weatherproof (6)
soothe (5)
there's
fathom

[h]: h, wh (20)

whom (4)	prohibit
humane	heave
hatred	harmony
mahogany	here's
heifer	
adhesive	
upheld	
hardy	
whose (3)	
hydrophobia	
hermit	

Nasals

[m]: m, mb, mm, mn (21)		
mercury	formerly	temperamentally
terminal	limbs	commerce
dimmer	comedy	remainder
crumbs	armory	
misty	mere	
condemn (6)	mimic	
transmission	famine	
mammal	movable	
administer	compromise	
mural	premium	
dramatically	amendment	
balm	sentimentally	

[n]: n, nn, kn, gn (22)		
noticeable	kin	enlarge
knuckle	cannibal	knocked
sovereign	anniversary	tonic
acknowledge	frantically	venison
fantasy	nasal	
innumerable	ninety-four	
knead	ban	
tenor	peninsula	
elimination	cinnamon	
feminine	eternal	
render	prominent	
penitentiary	sentiment	

[ŋ]: ng, n (23)	
rectangle	prolong
tongs	distinction
bungalow	puncture
gangster	linger
prong	settling
emptying	steadying
murmuring	strangle
upsetting	forgetting
displaying	transmitting
mingle	mangle
wasting	visiting
naming	gong

Glides 

[w]: w, wh (24)

weld	wary
whenever (4)	wed
forewarned	highways
warden	welfare
dwell	everywhere (2)
sewer	whim
wilt	swimmer
meanwhile (4)	
swoon	
await	
warp	

[r]: r, rr, wr, rh (25)

recognition	barrier	torrent
irritate	narrative	creed
wretched	parole	disarm
reside	warrant	bombard
pry	pyramid	unruly
rhythmically (6)	rash	artillery
uranium	enroll	
controlled	remnant	
starred	bracket	
embarrassment	apparatus	
cartridge	subterranean	

[ɔr] (subcategory of r): or, oor, our, ore, oar, ar, aur (26)

orbiting	gorilla	cordial
florist	swords	
award (5)	minority	
dinosaur (4)	folklore	
gourmet	organization	
sportsmanship	assortment	
corpse	disorder	
ornamental	coarse (5)	
supported	subordinate	
outdoors (3)	enforce	
formula	victorious	

[l]: l, ll, le, il, el, al, ell (27)

likeness	kiln	fragile
poll	landscape	stifle
privilege (6)	dimple	calculate
sullen	securely	mallet
satellite (5)	controllable	liable
lure	linoleum	relent
channel (5)	algebra	
galloping	translation	
alliance	mistletoe	
gargle	filter	
illustrate	cereal (5)	

[j]: y, i (28)

ammonia
yield (6)

Grade 7

Vowels

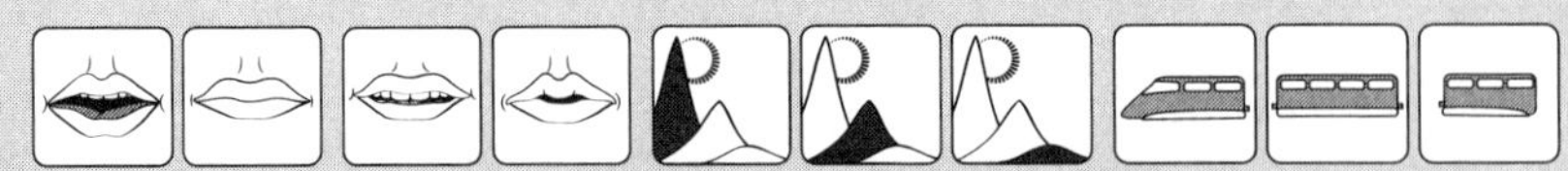

[i]: ee, e, y, e_e, ie, ea, ey, i, ei (29)		
neon	eternity	secrecy
pulley (5)	archeology	cavities
steeple	fiend	
believable	inferior	
compete	reek	
meteorite	conceive	
fee	precede	
alias	beseech	
bleak	repeal	
creative	allergy	

[ɪ]: i, y, ui (30)		
prescription	hypnotize	subscription
rebuild (4)	cynical	cavity
bristle	institution	synagogue
gymnastics	it's	
infinite	symptom	
criticism	destiny	
lisp	primitive	
administration	commit	
fiction	romantically	
civic	splint	

[ɛ]: ai, e, ea, ie, a, ae, hei (31)		
menu	tempest	succession
suspenders	millionaire	effective
unfriendly (4)	momentary (6)	respelled
heiress	heir	veteran
amend	perishable	welt
elevate	dependent	sentimental
preference	prospector	
recollection	realm	
pedestrian	assembly	
aerial	generator	
intellectual	detect	

[æ]: a, ai (32)		
prank	platinum	antler
gamble	tract	
accuracy	maniac	
romance	retract	
transplant	vat	
clan	sanitary	
plaid (5)	drastically	
sedan	supernatural	
tapestry	almanac	
gland	cancer	
atlas	fantastic	

[ɑ] & [ɔ]: a, o, al, ough, aw, augh, u, au, oa (33)		
monotonous	coffin	resolved
snob	gaunt	option
becalm	nausea	awning
lobster	abroad	squad
aeronautics	modify	assault
novelty	drama	
slaughter (6)	vaudeville	
automatically	sought (5)	
copied	falter	
bulb (3)	policy	

[ʊ]: oo, oul, u, ou (34)
plywood
boulevard
bulletin (6)
should've (6)
falsehood

Grade 7 Vowels *continued*

[u]: ue, o, ew, oo, oe, ui, ou, u_e, u, eu (35)

loot	groove	truce
resume	absolute	blue-eyed
detour	neutral (6)	
subdue	brutal	
approval (6)	overdue	
canoe (3)	brute	
proofread	pursuit	
noose	renew	
consume	fireproof	
coop	roost	
affluence	shrewd	
dual	brew	
typhoon	suicide	

[ɝ] & [ɚ]: er, or, ir, ear, ur, urr, irr, ure, ar, our, err (36)

fracture	survive	surge
laborer	geyser	quarterly
surplus	flirt	consumer
burlap	conferred (6)	turpentine
diaper	conveyor	culture
pastor	murder	dealer
squirrel (1)	waver	journeys
hearse	pursue	
turret	debtor	
regulator	burr	
entered	sculptor	
circulation	altered	
suitor	folder	

[ʌ] & [ə] : u, o, o_e, a, e, ou, oo, i (37)

comet	hub	sabotage
jumbo	societies	demerit
era	merit	fund
utmost	venom	slum
punctuate	municipal	overcome
riot	adaptable	bunt
correspondence	confront	sulk
miraculous	pendulum	traditionally
nun	subconscious	corruption
blood (3)	radius	utter
refund	saliva	chariot
redouble	abrupt	blunt
scum	reluctant	
erupt	tragedy	
abrupt	rivet	

Diphonemes

[eɪ] & [e]: ay, a, a_e, ai, ey, eigh, aigh, ea, ei (38)

reins (4)	patriotic (6)	sustain
survey (5)	rave	convey
bail	lubricate	
haste	delayed	
blockade	rayon	
donate (6)	basis	
stadium	rotation	
weighed	animated	
eightieth	staple	
regain	straighten (5)	
recitation	break (2)	
pastry	surveyed	

[oʊ] & [o]: o, o_e, oa, ow, oe, ou, ough, eau (39)

omen	elope	proprietor
vocabulary	thoroughbred	rodent
cantaloupe	trophy	chateau
domestic	gross	gallows
microscope	carload	afloat
alto	improvise	trousseau
minnow	donor	hoe (5)
thoroughly	microscopic	rotate
disown	smolder	tallow
bonus	beau	focus
donation	ownership	
cooperation	proposal	

[aʊ]: ou, ow, ough (40)

outrageous	counties
scour	
browse (6)	
trowel	
discount	
pout	
bough (5)	
outlet	
shroud	

[aɪ]: y, i, i_e, igh, ie, ai, ye, y_e, ui (41)

multiplied	hydrant	asylum
copyright	piloted	rivalry
rhyme (5)	lye	sightseeing
lifetime	anytime	sire
spry	nighttime	vile
guidance (6)	paralyze	agile
preside	stereotype	subscribe
despite	biography	defiance
aisle (5)	recital	

[oɪ]: oy, oi (42)

avoidance (6)
alloy
enjoying
adjoining
decoy (5)

[ju]: you, u_e, ew, ue, eau, u, uu (43)

unique	transfusion
persecute	unanimous
curfew (6)	speculate
you'll (3)	fumes
cue	universal
usage	beauty (3)
mutiny	utilize
vacuum (5)	united
masculine	muscular

Plosives

	[p]: p, pp	(1)
apparel	emancipation	precipitation
incompetent	cope	octopuses
protrude	gospel	catapult
deprive	implement	receptacle
epidemic	ample	repel
percolate	prosperous	interception
applied	oppress	separation
culprit	conspiracy	anticipate
repertory	temperate	purity
compliance	eclipse	displace
dependable	pamper	
piston	inspiration	
manipulation	perpendicular	
specify	chaplain	

	[b]: b, bb	(2)
throbbed	inability	celebrities
perturb	combustion	ballast
amphibious	liberal	
distribution	proverb	
habitation	navigable	
inhibit	admirable	
barbarian	drab	
prescribe	possibility	
lubricant	prefabricated	
bide	collaborate	
abbreviate	probability	
obtainable	superb	
embrace	blare	
probe	Sabbath	

	[t]: t, ed, tt	(3)
trend	immortality	dental
repetition	silhouette	restraint
intricate	antiseptic	sentinel
attentive	investment	partition
fatality	mentally	installation
detectable	consistent	testament
pertain	posterity	atone
wrapped (7)	penalty	contradict
optometry	gilt	pretense
editorial	photostat	criteria
battalion	tuition	
sultry	violate	

	[d]: d, dd, ed, t, tt	(4)
meddle (7)	deprived	reproduce
subcommittee	gardener	data
imbedded	antidote	heredity
duration	identity	tendency
concurred	calamity	boredom
vandal	macadam	adrift
clutter	doctrine	
recurred	memorandums	
submitting	verdict	
anatomy	radii	
dividend	blotter	
ordeal	remitted	

continues

[k]: k, c, ck, ch, cc, qu, que (5)		
halfback	clique	occupant
placard	scandal	curricula
technicality	monarchy	irk
likely	incorporate	recur
broker	chromium	synchronize
acclaim	chronology	choral
outskirts	lyrics	chronic
bankrupt	barnacle	masquerade
client	shirk	

[kw] (subcategory of k): qu (6)	
quota	aquariums
inquiry	equation
exquisite	quitting
consequently	acquittal
quintuplet	requisition
colloquial	sequel
qualms	
acquainted	
inquisitive	

[g]: g, gg, gu, gh, gue (7)		
embargo	haggard	pedigree
ignition	engagement	immigration
negative	aggravate	renegade
intrigue	integrate	saga
garlic	ghetto	guttural
aghast	congregation	legally
signify	morgue	
mortgage	guarantee (7)	
propaganda	emigrant	
brigade	pagan	

[ks]: x, xc, cc (8)	
exclusive	laxative
indexes	intoxicate
maximum	appendixes
acceptance (6)	expelled
extravagance	excerpt
reflex	
fixture	
exposition	
expectation	
annex	

[gz]: x (9)
exhibited
example (3)
exempt
exasperate
exertion
exemption

Sibilants

[s]: s, se, c, ss, ce, sc, z, ps (10)

depress
episode
waltz (6)
precipice
caress
masterpiece
accessory
consul
participation
incentive
fluorescent
dispense
susceptible
enforceable
massacre
adjacent
velocity
pseudonym
sect
intercept
lesson
embassy
assent
miscellaneous
censor
persistence
physics
discipline
psalms
insolent
salve
arson
sensitive
casserole
unstated
delicatessen
et cetera

[z]: s, z, se, zz, ze, ss (11)

revise
zeal
possession (4)
limousine
displease (7)
transit
galvanized
maze
plasma
analyze
visor
penalize
hazard
raising
quizzical
stanza
frenzy
mobilize
zest
capsize

[ʃ]: sh, ti, s, ch, ci, c, ssi, ss, si, xi (12)

chandelier
deception
pressure (5)
complexion (6)
depression
makeshift
chauvinist
classification
insurance (6)
relish
conditional
abolish
confidential
beneficial
ratio
dimension
aggression
facial
tarnish
cache
intermission
affliction
demolish
oceanographer (6)
petition
concession
earshot
militia
substantial
indication

[ʒ]: si, s, g, x (gʒ) (13)

enclosure
luxury (6)
anesthesia
fuselage
massage
casualty
provisional
invasion
collision

[tʃ]: ch, tch, t (14)

charter
spiritual
breach
stench
centuries
attachment
chime
virtue
char
scorch
legislature
chowder
dispatch

[dʒ]: j, dge, ge, g, gg, dg, d, dj (15)

graduation (6)
projection
fidget
jaunt
budget
suggestion (5)
logic
registration
clergy
dredge
eject
acreage
dislodge
indulge
shortage
juror
vengeance
legislation
discharge
negligence
prejudge
ledger
congenial
rejection
adjourn
virgin
surgeon
beverage
dejection

Fricatives

[f]: f, ff, gh, ph (16)		
forestry	fore	faculty
reinforce	siphon	prophecy
suffocate	graft	triumphantly
nymph	catastrophe	affirm
scaffold	benefited	fiancé
physicist	transform	decipher
clarify	offensive	specification
perfection	paraphrase	fiancée
enfold	amphibian	defect
modified	perforated	symphonic
stratosphere	pharmacy	confederate
roughly	philosophy	indefinite

[v]: ve, v, f (17)		
varsity	festivity	reverence
defective	divorce	evasive
of (1)	individually	sieve
lavatory	cavalry	revelation
evacuation	veranda	conservative
controversy	motive	videotape
inconvenience	observant	invert
derive	advocate	
pivot	perseverance	
envying	strive	
severely	privacy	
abbreviation	bravery	

[θ]: th (18)	
authentic	pathetic
thermos	apathy
synthesis	pathology
hypothesis	two-thirds
parenthesis	
anthem	
thesis	
authorize	
ruthless	
telepathy	

[ð]: th, the (19)
writhe
rhythmic
loathe
they're

[h]: h, wh (20)	
hectic	horde
hindrance	overhaul
mishap	alcoholic
who (1)	hospitable
hostage	whole (3)
comprehend	inhale
disinherit	left-handed
wholesome (6)	hilarious
hilt	hence
heroic	haphazard
homicide	

Nasals 

[m]: m, mb, mm, mn (21)

immunity	amiable	pantomime
proclamation	formality	prim
morality	cumbersome	intimate
demotion	inflammable	recommendation
plumbing	discriminate	encompass
contaminate	martyr	blemish
economy	memorandum	
condemn (6)	molest	
armament	transmit	
dominate	stamina	
numbed	omit	
emerald	commentator	

[n]: n, nn, kn, gn, pn (22)

acknowledge (7)	endorsement	bandanna
finance	perennial	denim
nomination	nautical	pollen
designer	flannel	reverend
phonics	connotation	tenement
sinister	invest	denotation
alignment	munitions	analyses
fanatic	annulled	
synopsis	concerned	
pneumonia	notorious	
gnawed	penalties	
din	convincible	

[ŋ]: ng, n (23)

sanctuary	omitting
oblong	pricing
blaming	brink
punctuality	rescuing
stating	aging
fitting	singe
confining	drunkard
dismaying	compelling
limiting	
function	
entangle	
timing	

Glides

[w]: w, wh (24)
whim (7)
subways
swindle
wily
swerve
elsewhere
weren't
swaying
wield
swivel

[r]: r, rr, wr, rh (25)		
wretched (7)	rhythm (6)	disrespect
relic	paralysis	rout
irritation	array	procedure
ardent	precise	rarity
corrode	retail	restrain
portray	barricade	oppression
quarreling	frolic	
remorse	reaction	
serenade	tremor	
delirious	enrich	
narration	bred	

[ɔr] (subcategory of r): or, oor, our, ore, oar, ar, aur (26)		
foremost	scoreboard	deformity
dinosaur (4)	conform	boarder
award (5)	endorse	corporal
torrid	resource	
resourceful	morsel	
hoard	implore	
passport	coordination	
adorn	distort	
courtship	portrait	
recorder	borne	
deport	abnormal	

[l]: l, ll, le, il, el, al, ell (27)		
leaflet	elapse	ailment
deluge	valet	reality
satellite (5)	evaluation	enrollment
galleries	floral	repellent
chronicle	accessible	fondle
allocate	foretell	subtle
rebels (6)	columnist	tallied
deflect	gill	dispelled
pitfall	incapable	
dislocate	collide	
civil (5)	multiple	

[j]: y, i (28)
vineyard
civilian
familiar (4)

Vowels

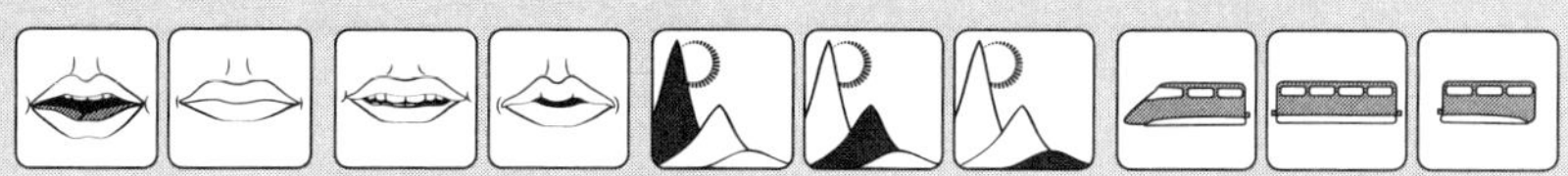

[i]: ee, e, y, e_e, ie, ea, ey, i, ei (29)

medleys	depleted	impeached
reprieve	meager	apprehend
diagnoses	agencies	morally
deceit	eaves	nominee
confetti	facilities	prosperity
appease	media	pulleys
genial	veneer	dingy
creation	lease	decelerate
recede	besiege	variation
locality	careen	artery

[ɪ]: i, y, ui (30)

flinch	facility	minimum
mystify	dictate	livid
significance	conspicuous	predicament
diminish	acquit	cistern
inmate	impressive	remit
lilt	pygmy	clinic
antonym	restrict	frill
definitively	systematic	specific
biscuit (6)	instill	lynch
eliminate	gymnasium	

[ɛ]: ai, e, ea, ie, a, ae, hei (31)

defenseless	identification	questionnaire
incense	congressional	edit
intersection	detest	steadfast
endeavor	breadth	petty
repent	inflection	indebted
contempt	agendas	
unfriendly (4)	lecturer	
aerial (7)	readiness	
paramedic	parish	
pedestal	impair	
heiress (7)	calendar	

[æ]: a, ai (32)

plaid (5)	chassis	matrimony
sanitation	morale	calorie
lapse	anesthetic	
pastel	pact	
overlap	diplomat	
static	drastic	
cascade	classic	
trance	rant	
enact	madly	
valor	slander	
abdicate	protagonist	

[ɑ] & [ɔ]: a, o, al, ough, aw, augh, u, au, oa (33)

faulty	dishonor	drawl
prophet	taunt	pauper
slaughter (6)	impart	haughty
optical	inoculate	periodical
coughed	gaudy	bondage
pawn	brawl	propulsion
fraud	molecule	pulpit
abroad (7)	inaugurate	communism
clog	sparse	calmly
brawn	installment	audio-visual
anonymous	auxiliary	

[ʊ]: oo, oul, u, ou (34)

livelihood
detour (6)
bulletin (6)
should've (6)

continues

[u]: ue, o, ew, oo, oe, ui, ou, u_e, u, eu (35)	
renewal	taboo
who's	multitude
recruit	intuition
canoe (3)	pollute
duplicate	aloof
plumage	removal
maroon	flue
rouge	dilute
elude	rheumatism
maneuver	illuminate
fortitude	cruelty
revenue	substitution
boon	lunatic

[ɝ] & [ɚ]: er, or, ir, ear, ur, urr, irr, ure, ar, our, err (36)		
prosper	verge	sponsor
immerse	hurdle	circuit
merge	cutlery	girder
squirrel (1)	impostor	rehearsal
nourish	encircle	scripture
surgery	binoculars	prior
confer	lottery	singular
inferred	assert	curdle
carburetor	molar	ambassador
curry	network	sterling
urgency	projector	dictator
adviser	survivor	orator
interpreter	perforate	currency

[ʌ] & [ə] : u, o, o_e, a, e, ou, oo, i (37)		
pious	mutilate	monastery
ventilate	economical	category
flood (4)	carousel	aristocrat
insulation	strenuous	reproduction
injustice	affirmative	escalator
overcome (7)	reduction	shun
ambiguous	versus	adequately
unison	alumni	musty
constable	brunt	
stadiums	corrupt	
rustic	debris	
toughness	fluster	
envelop	competent	
mason	justify	
recovery	trump	

Grade 8

Diphonemes

[eɪ] & [e]: ay, a, a_e, ai, ey, eigh, aigh, ea, ei (38)

haven	corporation	reins (4)
domination	straighten (5)	vocation
surveyed (7)	inflation	radium
inlaid	aide	actuate
bases	liberate	maim
patriotic (6)	conveyed	emigrate
misstated	evade	
vacancy	nominate	
defrayed	overlay	
allay	decade	
elated	refrain	
steak (2)	gait	
weighed (7)	consolation	
relocate	replaceable	

[oʊ] & [o]: o, o_e, oa, ow, oe, ou, ough, eau (39)

foe (6)	milestone	mode
nomad	abode	totally
vocal	lotion	provoke
grotto	molten	decompose
minnow (7)	global	
soprano	oases	
coma	veto	
portfolio	diploma	
furlough	bloated	
outgrowth	upholster	
poach	profile	
jovial	prone	
chateau (7)	corrosion	
maestro	potential	

[aʊ]: ou, ow, ough (40)

lounge	horsepower
penthouse	scoundrel
vouch	doubtless
profound	encounter
counterfeit	bout
drought (5)	outcast
endow	douse
renounce	flounder
devout	

[aɪ]: y, i, i_e, igh, ie, ai, ye, y_e, ui (41)

guidance (6)	lye (7)	piety
isolation	hibernation	defying
reliance	fortify	ply
naturalize	ignite	recycle
blight	plight	infantile
aisle (5)	rhymed	bias
identical	crises	rhyme (5)
dissatisfied	splice	certify
denial	pacify	

[oɪ]: oy, oi (42)

void
coy
foyer
recoil
ointment
buoyancy
decoys
convoy

[ju]: you, u_e, ew, ue, eau, u, uu (43)

interview (6)	infuriate
stimuli	hue (6)
immune	molecular
vacuums	fugitive
youth (4)	electrocute
beautiful (2)	prosecute
evacuate	communicable
commute	
arguing	

Plosives

[p]: p, pp (1)
suppress
comparable
optimistic
caption
supplement
disapprove
comply
expire
appraise
improbable

[b]: b, bb (2)
ebb
banister
deliberate
robust
abate
obscure
bosom
abominable
disband
obstinate

[t]: t, ed, tt (3)
contemplate
titanic
deceased
instantaneous
attribute
captivate
antagonize
undertone
smite

[d]: d, dd, ed, t, tt (4)
clutter (8)
imbedded (8)
redeem
endear
integrity
recurred (8)
celebrity
adhere
misconduct
expend

[k]: k, c, ck, ch, cc, qu, que (5)	
smock	intact
disembark	consecutive
mechanism	stricken
physique	à la carte
barracks	
masquerade (8)	
accomplice	
cauldron	
shank	

[kw] (subcategory of k): qu (6)
equivalent
squeamish
aquatic
vanquish
disqualify
quadruple

[g]: g, gg, gu, gh, gue (7)	
grimace	corrugated
regal	
aghast (8)	
sluggish	
gild	
bibliography	
insignificant	
guardian (6)	
aggressive	
intrigue (8)	

[ks]: x, xc, cc (8)
galaxy
expanse
acceptable (6)
expansion
extremity
complex
expenditure
excerpt (8)

[gz]: x (9)
exalt

Sibilants

[s]: s, se, c, ss, ce, sc, z, ps (10)

psychology (7)
cinema
stress
defensive
waltz (6)
sanity
amass
dispense (8)
disciple
indispensable
farce
siesta
descent (7)
asset
surpass
ecstasy

[z]: s, z, se, zz, ze, ss (11)

sarcasm
clause (7)
brazen
accusation
dispossess
transaction
analyze (8)
embezzle
desertion
disclose
cosmetics
disorganized
prism

[ʃ]: sh, ti, s, ch, ci, c, ssi, ss, si, xi (12)

insure (7)
sham
complexion (6)
judicial
issue (6)
malicious
aggression (8)
dimension (8)
compensation
famished
lavish
oceanographer (6)
champagne
abolition
specialist

[ʒ]: si, s, g, x (gʒ) (13)

illusion
conversion
casualty (8)
menagerie
diversion
regime
luxury (6)

[tʃ]: ch, tch, t (14)

urchin
chide
righteous
bewitch
detach
staunch
chafe
mutual

[dʒ]: j, dge, ge, g, gg, dg, d, dj (15)

pledge (5)
drudgery
ingenious
belligerent
dejected
suggest (4)
adjoin (8)
congestion
divulge
graduation (6)
strategy
intangible
agitate
disengage
converge

Fricatives

[f]: f, ff, gh, ph (16)	
suffice	deface
roughly (8)	pilfer
confiscate	
disfigure	
autobiography	
infantry	
doff	
superficial	
buffer	

[v]: ve, v, f (17)	
of (1)	evolution
verify	valise
canvass	ravage
avert	
sieve (8)	
livery	
divert	
conventional	
alcove	

[θ]: th (18)
swarthy
thwart
spendthrift

[ð]: th, the (19)
seethe (5)
heathen
noteworthy
breathe

[h]: h, wh (20)
who (1)
abhor
upheaval
whole (3)

Nasals

[m]: m, mb, mm, mn (21)

tempo	damnable
discomfort	
memorable	
numbed (8)	
remedial	
commodity	
aroma	
filament	
magistrate	

[n]: n, nn, kn, gn, pn (22)

pennant	discontent
artesian	inborn
felony	countenance
sovereign (7)	
disconnect	
pneumonia (8)	
analysis	
knead (7)	
terminate	

[ŋ]: ng, n (23)

- pending
- twinge
- flank
- foreboding

Glides

[w]: w, wh (24)

wane
overwhelm
wistful
dwindle
wean

[r]: r, rr, wr, rh (25)

wren (6)
irritable
arid
garrison
secrete
compress
rhythm (6)
replica
err
priority
filtration

[ɔr] (subcategory of r): or, oor, our, ore, oar, ar, aur (26)

hoard (8)
mortify
outdoors (3)
award (5)
dinosaur (4)
resource (8)
ornate
discord
formidable
encore
coronation

[l]: l, ll, le, il, el, al, ell (27)

lore
constellation
satellite (5)
debatable
delete
sublime
privilege (6)
imply
fillet
lament
allotment
labeled
dismantle
clerical

[j]: y, i (28)

billiards
rayon (7)

Vowels

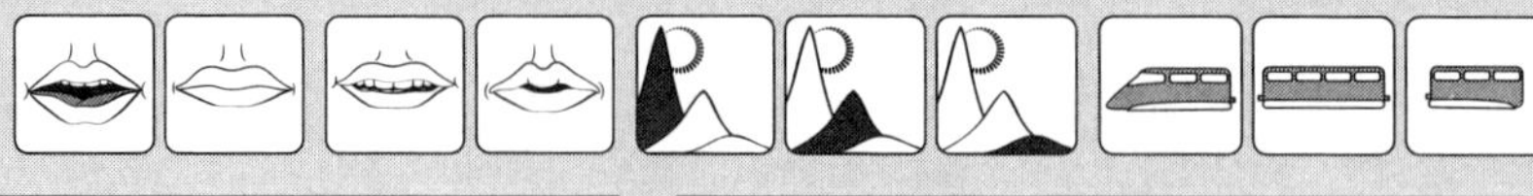

[i]: ee, e, y, e_e, ie, ea, ey, i, ei (29)

medieval
bleach
evolve
perceive
medleys (8)
dean
retrieve
absentee
surly
density
compulsory
insomnia
disbelief
serene
obsolete
grievance
concede
cleave
deem

[ɪ]: i, y, ui (30)

filtration
builder (4)
captivity
abyss
trivial
insignia
diction

[ɛ]: ai, e, ea, ie, a, ae, hei (31)

festive
unfriendly (4)
aerial (7)
meditate
telescope
benediction
adversary
parasite
veterinary
stealthy
millionaire (7)
heiress (7)

[æ]: a, ai (32)

landmark
reprimand
detract
nationality
adaptation
plaid (5)

[ɑ] & [ɔ]: a, o, al, ough, aw, augh, u, au, oa (33)

repulsive
slaughter (6)
homage
dauntless
embalm
sought (5)
pawn (8)
denomination
impulsive
abroad (7)
tolerate
arbitrary
retard
audition
fraternity

[ʊ]: oo, oul, u, ou (34)

hoodwink
contour
bully (5)
would've (6)

[u]: ue, o, ew, oo, oe, ui, ou, u_e, u, eu (35)	
revenue (8)	booty
platoon	reprove
brew (7)	maneuver (8)
recruit (8)	boulevard (7)
continuity	
superfluous	
luminous	
canoe (3)	
removal (8)	
duly	
destitute	

[ɝ] & [ɚ]: er, or, ir, ear, ur, urr, irr, ure, ar, our, err (36)	
inferred (8)	circumstance
spurn	slur
vendor	adverse
caper	earthy
squirrel (1)	texture
diverge	turret (7)
glamour	refinery
wayward	ferment
inertia	
infirmary	
rapture	

[ʌ] & [ə] : u, o, o_e, a, e, ou, oo, i (37)	
tantrum	escapade
entrust	snub
toughness (8)	cult
confront (7)	disrupt
arrogant	isthmus
deplore	flood (4)
conscientious	gruesome
legitimate	lust
category (8)	caliber
random	porous

Diphonemes

[eɪ] & [e]: ay, a, a_e, ai, ey, eigh, aigh, ea, ei (38)	
fumigate	break (2)
distasteful	freight (4)
patriotic (6)	straighten (5)
bayonet	gaiety
surveyed (7)	avail
ordain	originate
amputate	fray
conveyed (8)	crave
beige	baste
vacate	feint

[oʊ] & [o]: o, o_e, oa, ow, oe, ou, ough, eau (39)	
outgrowth (8)	diplomacy
coincide	domain
beau (7)	foe (6)
thorough (7)	
smote	
probation	
cantaloupe (7)	
anecdote	
bestow	
reproach	

[aʊ]: ou, ow, ough (40)	
rebound	renown
counteract	dowry
overpower	cower
abound	
spouse	
denounce	
foundry	
bough (5)	

[aɪ]: y, i, i_e, igh, ie, ai, ye, y_e, ui (41)	
rhyme (5)	rye (6)
insight	misguided
glorify	decisive
occupies (6)	vial
concise	dire
tripod	amplify
foresight	entice
aisle (5)	

[oɪ]: oy, oi (42)
turmoil
envoy
loiter

[ju]: you, u_e, ew, ue, eau, u, uu, eu (43)	
youth (4)	mute
vacuum (5)	articulate
compute	
discontinue	
impudent	
vestibule	
beauty (3)	
pewter	
feudal	

Plosives

[p]: p, pp (1)

perspective
spontaneous
comparative
appropriation
imperative
misconception
recuperate
compile
piteous
displacement
hapless
passive
dispel
imperial

[b]: b, bb (2)

tributary
reimburse
abbreviate (8)
intolerable
vertebrate
laborious
benevolent
stability
liability
befall
bevy
embellish
soluble

[t]: t, ed, tt (3)

reinstate
clientele
deceased (9)
glutton
scrutinize
vitality
maternal
consultant
buttress
invertebrate
stolid
devastate
petite
frustrate
remittance
impertinent
simultaneous
debutante
tantalize
pertinent

[d]: d, dd, ed, t, tt (4)

sundry
demure
bereaved
ratify
sodden
adept
solitude
deft
indisposed
credentials
sediment
intensity
providence
subcommittee (8)
despondent
eradicate

[k]: k, c, ck, ch, cc, qu, que (5)

preoccupy
opaque
lacquer (7)
stricken (9)
encampment
unkempt
parochial
nucleus
skeptical
oblique
grotesque
pictorial
transcribe
capillary
havoc
skirmish
invoke
disconcert
secluded

[kw] (subcategory of k): qu (6)

quell
liquidate
eloquent
quandary

continues

[g]: g, gg, gu, gh, gue (7)	
legacy	pigment
dialogue (7)	impregnate
segregate	aggressive (10)
organic	gala
dogged	designate
ghetto (8)	bogus
stigma	
gore	
dregs	
guess (4)	

[ks]: x, xc, cc (8)
excitable (7)
apex
exponent
accentuate
expressly

[gz]: x (9)
exotic
exuberant

Sibilants

[s]: s, se, c, ss, ce, sc, z, ps (10)

sterile
listless
incandescent
concept
diverse
psychiatrist
farce (10)
centennial
waltz (6)
assess
incite
discretion
passé
emboss
convalescent
sorority
dispensary
sear

[z]: s, z, se, zz, ze, ss (11)

ooze (6)
prose
organism
dispossess (9)
feasible
reorganize
materialize
spasm
plausible
cosmic
embezzle (9)
displease (7)
pulverize
glaze
economize

[ʃ]: sh, ti, s, ch, ci, c, ssi, ss, si, xi (12)

negotiate
ocean (3)
regression (7)
impoverish
chagrin
pressure (5)
atrocious
nonchalant
dimension (8)
ravish
insurance (6)
perception
automation
obnoxious
ashtray
demolition
crucial
proficient
fictitious
impartial

[ʒ]: si, s, g, x (gʒ) (13)

luxury (6)
enclosure (8)
compulsion
menagerie (9)
evasion

[tʃ]: ch, tch, t (14)

impeach
ritual
dispatch (8)
virtual
infatuate
obituary

[dʒ]: j, dge, ge, g, gg, dg, d, dj (15)

jargon
vigil
frontage
suggestion (5)
pilgrimage
germinate
illegitimate
tangent
collegiate
readjust
porridge (6)
graduation (6)
drudgery (9)

Fricatives

[f]: f, ff, gh, ph (16)

trough (5)
deficient
flaunt
affront
phenomenal
insufferable
diaphragm
rebuff
flange
reformatory
focal

[v]: ve, v, f (17)

provincial
overture
evaluate
bevel
vigilant
grovel
inevitable
carnivorous
naïve
of (1)
prevalent
invincible
revel
frivolous
delve

[θ]: th (18)

enthrall
slothful
theorem
ethical
amphitheater
froth
synthetic

[ð]: th, the (19)

betroth
seethe (5)

[h]: h, wh (20)

whole (3)
horoscope
hybrid
rehabilitate
enhance
humiliate
hypocrite
who (1)
hovel

Nasals

[m]: m, mb, mm, mn (21)

demented
humility
scrimmage
animation
solemn (5)
crumbs (7)
emit
dilemma
mire
rostrum
complement
smelt
malady

[n]: n, nn, kn, gn, pn (22)

gnawed (8)
infernal
annihilate
knuckle (7)
notary
pneumonia (8)
artisan
inclination
renovate
stagnant
finale
ordinance
pinnacle
component
pregnant
annul
dominant

[ŋ]: ng, n (23)

inkling
scavenger
elongate
unerring
invigorating

Glides

[w]: w, wh, u (24)

whet
waif
stalwart
suave
wan
wend

[r]: r, rr, wr, rh (25)

rhapsody	predominant
narrate	transpose
wretched (7)	reconcile
brandish	aura
radical	incriminate
granulate	curriculum
disreputable	ferret
strata	enrapture
parry	

[ɔr] (subcategory of r): or, oor, our, ore, oar, ar, aur (26)

horticulture	formulate
portal	boarder (8)
dinosaur (4)	morbid
award (5)	foreshorten
coordination (8)	coroner
oratory	
courtship (8)	
dormant	
demoralize	

[l]: l, ll, le, il, el, al, ell (27)

diligent	flail	solvent
guerrilla	ensemble	cull
civil (5)	illustrious	
inaccessible	fallacy	
labeled (9)	bland	
reputable	lateral	
droll	allure	
tolerable	sally	
satellite (5)	cuticle	

[j]: y, i (28)

billiards (9)
vineyard (8)

Vowels

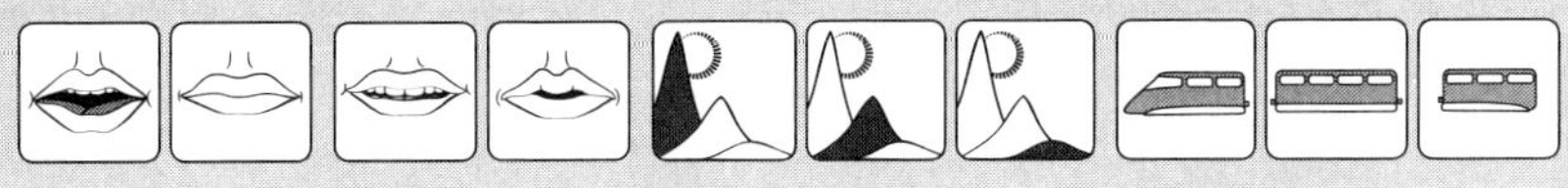

[i]: ee, e, y, e_e, ie, ea, ey, i, ei (29)

precinct
esteem
medley
tier
depreciate
perceive (9)
colleague
genteel
maternity
eccentric
obscene
embryo
portly
glean
reverie
deity

[ɪ]: i, y, ui (30)

rebuild (4)
symmetry
glib
pessimistic
dignitary
rift
infidel
mystical
oblivion
crypt
spinster
inducement
cryptic
hysteria

[ɛ]: ai, e, ea, ie, a, ae, hei (31)

rend
carat
aerial (7)
necessitate
stealthy (9)
temperance
heirloom
marital
debit
pensive
sector
friendly (2)
contemptible
impair (8)

[æ]: a, ai (32)

emancipate
tabulate
plaid (5)
banter
satire
ravenous
panorama
magnitude
stance
contraband

[ɑ] & [ɔ]: a, o, al, ough, aw, augh, u, au, oa (33)

maul
appalling
tart
ultimatum
convulsion
fraught
somber
default
qualm
intolerant
tawdry
abroad (7)
roster
bauble
consolidate
ardor
coughed (8)

[ʊ]: oo, oul, u, ou (34)

livelihood (8)
contour (9)
could've (6)
bully (5)

continues

[u]: ue, o, ew, oo, oe, ui, ou, u_e, u, eu (35)

rue	uncouth
prudent	inclusive
removal (8)	rheumatism (8)
canoe (3)	platoon (9)
hew	strewn
pursuit (7)	dupe
disillusion	induce
stupefy	
conclusive	
prelude	
frugal	

[ɝ] & [ɚ]: er, or, ir, ear, ur, urr, irr, ure, ar, our, err (36)

turret (7)	inert
dirge	thermal
linear	enumerate
benefactor	revert
gender	curvature
stupor	squirrel (1)
earthy (9)	furtive
urban	cater
saturate	inferred (8)
courier	girth
deteriorate	manor

[ʌ] & [ə] : u, o, o_e, a, e, ou, oo, i (37)

gruesome (9)	heritage	superstructure
monologue	gratify	defile
rupture	tumult	humbug
blood (3)	duct	
recovery (8)	enunciate	
axiom	redouble (7)	
equilibrium	idiom	
callous	humdrum	
forum	bedlam	
dialect	gaseous	

Diphonemes

[eɪ] & [e]: ay, a, a_e, ai, ey, eigh, aigh, ea, ei (38)	
manipulate	mainstay
feign	taint
curtail	straighten (5)
steak (2)	dilate
namesake	patron
cancellation	radiate
surveyed (7)	belated
disintegrate	deflation
vacate (9)	conveyance
freight (4)	bane

[oʊ] & [o]: o, o_e, oa, ow, oe, ou, ough, eau (39)	
poultry (6)	dote
superimpose	profess
bestow (9)	bolster
mediocre	dough (5)
disown (7)	hoax
scope	momentum
hoe (5)	folio
beau (7)	
revoke	
torso	

[aʊ]: ou, ow, ough (40)
dowry (9)
renown (9)
drought (5)
abound (9)
confound

[aɪ]: y, i, i_e, igh, ie, ai, ye, y_e, ui (41)		
rectify	irony	dynamic
misguided (9)	aisle (5)	
belie	canine	
timely	pliable	
geyser (7)	hydraulic	
rye (6)	rhyme (5)	
patronize	enlighten	
ire	subside	

[oɪ]: oy, oi (42)
convoy (8)
exploit
foible
boisterous

[ju]: you, u_e, ew, ue, eau, u, uu, eu (43)	
unify	stimulate
vacuums (8)	pewter (9)
youth (4)	cumulative
putrid	futile
beautiful (2)	disuse
feudal (9)	
mutation	
you (K)	
revue	

Plosives

[p]: p, pp (1)	
pallor	preposterous
decrepit	implicate
supple	replenish
inseparable	capitalist
rampant	discrepancy
partisan	
epic	
tepid	
populace	
opportune	
intersperse	

[b]: b, bb (2)	
encumbered	invariably
rabble	lobbyist
semblance	rebate
gibe	obliterate
abrasion	subsistence
quibble	fabricate
vibrant	
habitable	
bier	
inhibition	
palatable	

[t]: t, ed, tt (3)	
entail	retaliate
retina	inadvertently
constitute	statistics
attaché	stilt
trite	fastidious
detriment	instigate
wrapped (7)	vintage
tentative	intonation
intermittent	tenure
nocturnal	palpitate
typify	

[d]: d, dd, ed, t, tt (4)	
disheveled	sedative
condolence	candid
fidelity	
addict	
mettle	
indict	
paradox	
fetter	
dilapidated	
idolatry	
eddy	

[k]: k, c, ck, ch, cc, qu, que (5)	
masquerade (8)	kinetic
evoke	raucous
precarious	accost
brusque	kindred
ransack	concussion
occult	traction
ascribe	caricature
scourge	incur
catechism	

[kw] (subcategory of k): qu (6)	
tranquil	query
quadrant	
inquest	
requisite	
equity	
squander	
bequeath	
antiquated	
inquisition	

[g]: g, gg, gu, gh, gue (7)	
malignant	flagrant
egotism	legation
gauntlet	engross
vagrant	prodigal
dogged (10)	transgress
centrifugal	incongruous
vogue	
guide (2)	
integral	
aghast (8)	

[ks]: x, xc, cc (8)	
dexterity	context
influx	
affix	
explicit	
accentuate (10)	
convex	
extortion	
excitable (7)	
proxy	
laxity	

[gz]: x (9)
nonexistent
exhilarated
exultation

Sibilants

[s]: s, se, c, ss, ce, sc, z, ps (10)

pacifist
saunter
condescend
incendiary
essence
assail
discern
impassive
waltz (6)
abscess
infest
subsequent
onslaught
psychosis
cessation
ascertain
reminiscent
disperse
solicit
placid
insensible

[z]: s, z, se, zz, ze, ss (11)

zodiac
quizzical (8)
renaissance
surmise
generalize
mausoleum
analyze (8)
guise
cosmopolitan
zenith
resolute
repose
infuse
composite
jeopardize
desirous

[ʃ]: sh, ti, s, ch, ci, c, ssi, ss, si, xi (12)

issue (6)
abashed
dimension (8)
ricochet
restoration
emaciated
socialist
compassion
burnish
initiative
disproportionate
judicious
obnoxious (10)
beneficiary
faction
nationalism
relinquish
oceanographer (6)
obsession
sensual
conception

[ʒ]: si, s, g, x (gʒ) (13)

regime (9)
composure
intrusion
luxuriant
revulsion
fissure
expulsion
aversion

[tʃ]: ch, tch, t (14)

virtuoso
presumptuous
bewitch (9)
voucher
franchise

[dʒ]: j, dge, ge, g, gg, dg, d, dj (15)

algae
jaded
graduation (6)
regent
budget (8)
vantage
jocular
congeal
interject
suggest (4)
derange
jurisdiction
lineage
patronage
rejuvenate
abridge
adjoin (8)
elegy

Fricatives

[f]: f, ff, gh, ph (16)

infer
roughly (8)
asphyxiation
affable
profuse
confederation
manifold
differential
infamous
epitaph
affectation
philanthropy
emphatic
ramification
defer

[v]: ve, v, f (17)

vanguard
concave
servitude
vivacious
cavalcade
versatile
traverse
of (1)
primeval
respective
devious
severance
intervene
virile
fervent
inverse
covet

[θ]: th (18)

labyrinth
therapeutic
atheist
lethal
pathos
aftermath

[ð]: th, the (19)

wither (5)
swathe
feather (2)

[h]: h, wh (20)

hulk
whole (3)
hallucination
prohibition
who (1)
inherent
habitat

Grade 11

Nasals

[m]: m, mb, mm, mn (21)		
permeate	simulate	claimant
immaterial	noncommittal	
memoir	mien	
promenade	intimidate	
damnable (9)	materialistic	
commemorate	commentary	
numbed (8)	premonition	
impediment	dissimilar	
remuneration	immaculate	

[n]: n, nn, kn, gn, pn (22)	
pneumonia (8)	mannerism
intern	eminent
knoll	inanimate
biennial	align
malign	nominal
resonance	personnel
financier	annals
concur	reconnaissance
innovation	inane

[ŋ]: ng, n (23)
impending
tincture
bilingual
excruciating
infringe

Glides

[w]: w, wh, u (24)

whimsical
linguistic
cuisine
wanton
suede
waiver

[r]: r, rr, wr, rh (25)

recurrence
realist
awry
truism
rhapsody (10)
drone
resurrection
wrangle
insurrection
bizarre
prostrate
increment
reciprocate
acrid
erratic
irrepressible
erroneous
criterion

[ɔr] (subcategory of r): or, oor, our, ore, oar, ar, aur (26)

dinosaur (4)
coordination (8)
recourse
hoard (8)
award (5)
contortion
encore (9)
forage
discourse
mentor
orgy
disgorge

[l]: l, ll, le, il, el, al, ell (27)

malleable
privilege (6)
libel
satellite (5)
elliptical
solace
incredulous
invaluable
alleviate
enliven
negligible
oscillate
populous
alienate
resplendent
stimulus
infallible
gullible

[j]: y, i (28)

lenient
rayon (7)

Grade 11

Vowels

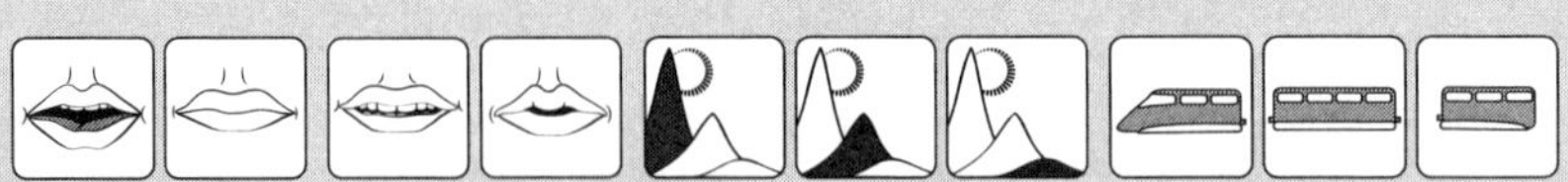

[i]: ee, e, y, e_e, ie, ea, ey, i, ei (29)

orient	meander	myriad
medley (10)	imagery	vehement
veer	menial	analogy
intercede	reverie (10)	
penal	dissociate	
misdemeanor	feline	
idealism	peerless	
sensory	treatise	
deceit (8)	secede	
deviate	deplete	

[ɪ]: i, y, ui (30)

syndicate	biscuit (6)
depict	tryst
pillage	polygamy
stint	retribution
acquisition	simile
facilitate	imperil
physiology	cyst
assimilate	indignity
infidelity	
limpid	

[ɛ]: ai, e, ea, ie, a, ae, hei (31)

debonair	inept	confection
forbearance	projectile	contend
irrelevant	overbearing	
parable	manifest	
repress	monetary	
retention	celestial	
discredit	variance	
unfriendly (4)	aerial (7)	
heirloom (10)	culinary	

[æ]: a, ai (32)

valid	sanction
cataract	interaction
plaid (5)	rabid
transition	
malice	
incapacitate	
status	
mandatory	
facet	

[ɑ] & [ɔ]: a, o, al, ough, aw, augh, u, au, oa (33)

qualm (10)	neurotic
concoct	pompous
sought (5)	fraudulent
tawdry (10)	marauder
distraught	slovenly
audit	arbitration
adulterated	stark
broaden (6)	taut
preponderance	botany
metropolis	ulterior

[ʊ]: oo, oul, u, ou (34)

contour (9)
bulwark
would've (6)
plywood (7)

[u]: ue, o, ew, oo, oe, ui, ou, u_e, u, eu (35)	
nutrient	dubious
recruit (8)	reproof
amplitude	boudoir
canoe (3)	ruse
lewd	ensue
scruple	interlude
reprove (9)	
impromptu	
translucent	
neuter	
peruse	

[ɝ] & [ɚ]: er, or, ir, ear, ur, urr, irr, ure, ar, our, err (36)	
rehearsal (8)	perjury
courier (10)	burlesque
purge	terse
literal	conferred (6)
deter	rigorous
aspiration	burlesque
squirrel (1)	interloper
occurring (6)	converse
successor	aperture
fervor	lunar
posterior	

[ʌ] & [ə] : u, o, o_e, a, e, ou, oo, i (37)		
glut	fluctuate	derelict
informant	antecedent	perpetrate
lustrous	pugnacious	interrogate
prospectus	conductive	redundant
redouble (7)	instrumental	contemptuous
collective	alimony	monotone
income (6)	confront (7)	
adage	maladjustment	
momentous	ominous	
flood (4)	condiment	

Diphonemes

[eɪ] & [e]: ay, a, a_e, ai, ey, eigh, aigh, ea, ei (38)

feign (10)	straighten (5)	natal
liaison	mania	dilate (10)
parley	bayonet (9)	conveyance (10)
freight (4)	abstain	impale
surveyed (7)	orientation	
notation	sedate	
disdainful	consecrate	
savory	pollinate	
profane	complacent	
break (2)	patriotic (6)	

[oʊ] & [o]: o, o_e, oa, ow, oe, ou, ough, eau (39)

memento	procure	scapegoat
hoe (5)	condone	
doleful	motif	
furlough (8)	potent	
outgrowth (8)	chateau (7)	
cantaloupe (7)	gusto	
fallow	propriety	
broach	rote	
odious	morose	
invocation	diagnosis	

[aʊ]: ou, ow, ough (40)

drought (5)
surmount
glower
unfounded
carouse

[aɪ]: y, i, i_e, igh, ie, ai, ye, y_e, ui (41)

dynasty	sobriety	rite
vie	lye (7)	
geyser (7)	nullify	
reprisal	enlighten (10)	
guidance (6)	rhyme (5)	
docile	binder	
transpire	cite	
aisle (5)	tirade	

[oɪ]: oy, oi (42)

invoice
boycott
devoid

[ju]: you, u_e, ew, ue, eau, u, uu, eu (43)

stipulate	commune
you've (3)	pugilist
beauty (3)	you'll (3)
muse	feudal (9)
vacuum (5)	refute
askew	
insinuate	
revue (10)	
diffuse	

Plosives

[p]: p, pp (1)	
prodigious	precept
supplant	flippant
dissipate	impasse
conscription	recapitulate
depot	impotent
parlance	propitious
apparition	
expatriate	
complicity	
nuptial	
trepidation	

[b]: b, bb (2)	
bureaucracy	upbraid
quibble (11)	nebulous
voluble	debase
banal	
macabre	
aberration	
habituate	
imbibe	
obesity	
amicable	
reprobate	

[t]: t, ed, tt (3)	
pirouette	nonentity
restitution	altruistic
tome	pittance
latent	tremulous
autocracy	repartee
pretentious	distend
detonate	solstice
ensconced	taciturn
intrepid	impetus
travesty	desultory
militant	extricate

[d]: d, dd, ed, t, tt (4)	
dissonant	decadence
longevity	supersede
sodden (10)	adamant
advent	levity
indiscretion	edict
vindictive	
depraved	
mettle (11)	
ponderous	
fetid	
mediation	

[k]: k, c, ck, ch, cc, qu, que (5)	
pique	impeccable
rancor	anarchy
hackneyed	preclude
acrimonious	caustic
askance	succulent
epoch	piquant
ecclesiastical	
carnage	
acclimate	

[kw] (subcategory of k): qu (6)
obsequious
quintessence
colloquial
equestrian
squalid
inequity

continues

[g]: g, gg, gu, gh, gue (7)	
paragon	augment
demagogue	pragmatic
figuratively	
aggregate	
regale	
gregarious	
ghetto (8)	
propagate	
guardian (6)	
dogmatic	

[ks]: x, xc, cc (8)
fixation
pretext
extradition
acceptable (6)
exodus
proximity
excerpt (8)
maxim
exorbitant
inexplicable

[gz]: x (9)
exhort
exude

Sibilants

[s]: s, se, c, ss, ce, sc, z, ps (10)

redress
nuance
consensus
psyche
idiosyncrasy
connoisseur
effervescence
subsidiary
finesse
transcend
incidence
missive
waltz (6)
perverse
agnostic
quiescent
promissory
edifice
reticence
crass
sumptuous
felicity
desist

[z]: s, z, se, zz, ze, ss (11)

intrinsic
ostracize
symposium
analyze (8)
raze
embezzle (9)
repository
visage
plagiarize
renaissance (11)
prosaic
clause (7)
presentiment
resilient
comprise
transient
metabolism

[ʃ]: sh, ti, s, ch, ci, c, ssi, ss, si, xi (12)

garish
echelon
pension (7)
censure
complexion (6)
officiate
charlatan
volition
pressure (5)
digression
tenacious
ostentatious
patrician
gauche
ocean (3)
avocation
differentiate
facetious

[ʒ]: si, s, g, x (gʒ) (13)

composure (11)
bourgeois
fusion
luxuriant (11)
protégé
delusion
aspersion

[tʃ]: ch, tch, t (14)

chastise
impetuous
satchel (7)
postulate
capitulate
chattel
stature
petulant
constituent

[dʒ]: j, dge, ge, g, gg, dg, d, dj (15)

juxtaposition
graduation (6)
suggestion (5)
tutelage
conjecture
abridge (11)
drudgery (12)
marginal
effigy
abject
incorrigible
vestige
adjoin (8)
prodigy
resurgent
subjugate
cogitate
trajectory
allege
degenerate

Fricatives

[f]: f, ff, gh, ph (16)

confidant
metamorphosis
affiliate
roughly (8)
amorphous
suffused
atrophy
ineffectual
phlegmatic
topography
orifice
diffident
facsimile
affluent
infinitesimal

[v]: ve, v, f (17)

avid
of (1)
venerate
evince
absolve
relevant
omnivorous
novice
impervious
inveterate
prevaricate
conversant
reverberation
voluptuous
derivative

[θ]: th (18)

ethereal
orthodox
aesthetic
lethargy
pathological
anathema

[ð]: th, the (19)

tithe
father (K)
scathing
lithe

[h]: h, wh (20)

heterogeneous
whole (3)
adherent
heretic
who (1)
heresy
harangue

Nasals

[m]: m, mb, mm, mn (21)	
mercenary	succumb
impel	pandemonium
plummet	homogeneous
amalgamate	communal
solemn (5)	referendum
disseminate	somnambulist
ruminate	extemporaneous
acumen	posthumous
indemnity	imminent

[n]: n, nn, kn, gn, pn (22)	
align (11)	innate
inundate	sustenance
garnish	connive
animosity	omnipotent
knead (7)	penitent
pneumatic	abstinence
somnolent	sedentary
inalienable	indeterminate
annuity	dissent

[ŋ]: ng, n (23)
defunct
scintillating
perfunctory
astringent
sanctify
juncture
cantankerous

Glides

[w]: w, wh, u (24)

languid
languish
dissuade
whimsical (11)
unwitting

[r]: r, rr, wr, rh (25)

rhetoric
pariah
respite
iridescent
wrest
estrange
harass
lucrative
bizarre (11)
raiment
wreak
deride
arable
credulous

[ɔr] (subcategory of r): or, oor, our, ore, oar, ar, aur (26)

sordid
outdoors (3)
award (5)
inordinate
scoreboard (8)
concourse
tortuous
foray
meritorious
predatory
corollary
dinosaur (4)
transitory
foreshorten (10)

[l]: l, ll, le, il, el, al, ell (27)

illicit
lurid
civil (5)
sully
correlate
collateral
amenable
satellite (5)
indolent
intelligentsia
pallid
manacle
allegory
pestilence
blasphemy
solicitous
despicable

[j]: y, i (28)

vineyard (8)
pinion
paraphernalia

Grade 12

Vowels 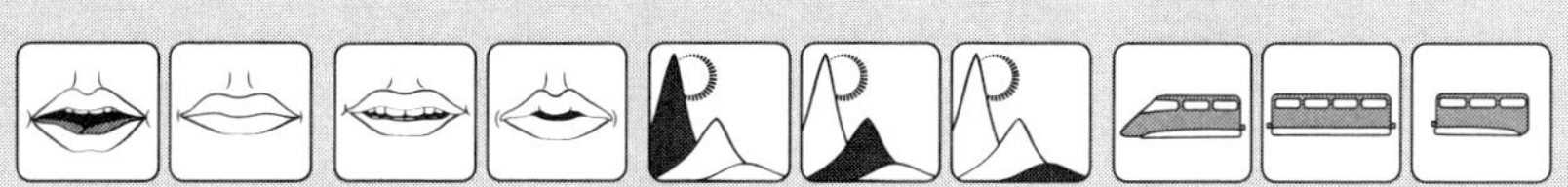

[i]: ee, e, y, e_e, ie, ea, ey, i, ei (29)

salient
impede
demeanor
plebeian
motley
cerebral
retrieve (9)
convene
discreet
austerity
expedient
remiss
infamy
median
sagacity
deify
cede
replete
entreat
conciliate
amity

[ɪ]: i, y, ui (30)

sylvan
figment
vindicate
rebuild (4)
definitive
recipient
prolific
fiscal
nondescript
insidious
supercilious
hieroglyphics
affinity
recrimination
inimitable
diminutive

[ɛ]: ai, e, ea, ie, a, ae, hei (31)

deference
retrospect
tenet
zealous
maritime
aerial (7)
friendly (2)
preeminent
recompense
amenity
infectious
debonair (10)
salutary
precedent
regenerate
electorate
incessant
penance
heiress (7)
proletariat
disparage
concentric

[æ]: a, ai (32)

panacea
protract
vapid
plaid (5)
sacrilege
caste
infraction
tractable
platitude
procrastinate
invalidate
refraction
abstraction
indefatigable
gratis

[ɑ] & [ɔ]: a, o, al, ough, aw, augh, u, au, oa (33)

laudable
vulnerable
embody
pawn (8)
plaudit
fraught (10)
quaff
sardonic
coughed (8)
gossamer
indomitable
larceny
abroad (7)
modulate
omnibus
admonish
embalm (9)
prerogative
audacity
culminate
arduous

[ʊ]: oo, oul, u, ou (34)

bulwark (10)
detour (6)
hoodwink (9)
could've (6)

continues

[u]: ue, o, ew, oo, oe, ui, ou, u_e, u, eu (35)	
pseudo	canoe (3)
ludicrous	plurality
removal (8)	accrue
obtuse	recluse
festoon	delude
lewd (10)	moot
voluminous	rudiment
allude	statute
uncouth	pursuit (7)
duplicity	astute
lieu	

[ɝ] & [ɚ]: er, or, ir, ear, ur, urr, irr, ure, ar, our, err (36)		
overt	interminable	dissertation
nourish (8)	elixir	vernacular
gird	tertiary	subservient
surreptitious	inferred (8)	pursuant
verdant	herbivorous	duress
modular	nurture	usurp
predecessor	squirrel (1)	coerce
garner	pastoral	regurgitate
circumvent	candor	nomenclature
hearse (7)	firmament	
reiterate	secular	

[ʌ] & [ə] : u, o, o_e, a, e, ou, oo, i (37)		
toughness (8)	pungent	veritable
subsidy	caucus	repugnant
incumbent	flood (4)	bigot
parody	disconsolate	cadaverous
insipid	chasten	dub
comely	ignominious	effrontery
gamut	abduct	
agrarian	mundane	
supplication	affidavit	
abet	promontory	

Diphonemes

[eɪ] & [e]: ay, a, a_e, ai, ey, eigh, aigh, ea, ei (38)

steak (2)	chaste	relegate
pervade	plaintive	gradation
magnate	straighten (5)	ingratiate
constraint	defray	delineation
insatiable	surveillance	conclave
surveyed (7)	baleful	
obviate	staid	
emanate	vacillate	
sleigh (3)	abeyance	
inveigle	reparation	

[oʊ] & [o]: o, o_e, oa, ow, oe, ou, ough, eau (39)

fiasco	incognito	hallowed
poultry (6)	protocol	precocious
beau (7)	foe (6)	
erode	stoic	
limbo	vociferous	
sojourn	denote	
hoax (10)	pedagogy	
ignoble	doughnut (5)	
misnomer	disown (7)	
extol	cohort	

[aʊ]: ou, ow, ough (40)

- prowess
- tantamount
- bough (5)
- expound
- recount
- propound

[aɪ]: y, i, i_e, igh, ie, ai, ye, y_e, ui (41)

misguided (9)	vie (10)	scion
rhyme (5)	notoriety	contrive
vicarious	decry	privation
personify	citation	rye (6)
beguile	prototype	domicile
senile	foresight (9)	
aisle (5)	ribald	
guile	volatile	

[oɪ]: oy, oi (42)

- boycott (10)
- poignant
- adroit
- embroiled

[ju]: you, u_e, ew, ue, eau, u, uu, eu (43)

utilitarian	beautiful (2)
repudiate	imbue
feudal (9)	subterfuge
you (K)	promiscuous
meticulous	ingenuous
sinuous	you'll (3)
askew (11)	innuendo
opulent	
vacuum (5)	

Plosives

[p]: p, pp (1)

spurious
propensity
empirical
opprobrious
perspicacious
impious
caprice
terpsichorean
apprise
prescience
catalepsy
puerile
asperity
palpable
epicure

[b]: b, bb (2)

umbrage
debacle
inscrutable
shibboleth
adumbrate
glabrous
salubrious
lugubrious
probity
equable
obdurate
bombast

[t]: t, ed, tt (3)

martinet
pristine
tacit
contrite
attest
fortuitous
ensconced (12)
turgid
blatant
dilettante
ostensible
peripatetic
intemperate
recalcitrant
epitome
extant

[d]: d, dd, ed, t, tt (4)

pedantic
eddy (11)
inured
antipodes
clutter (8)
invidious
undulate
disparity
enamored
strident
maudlin
indigent
adjure
declamation

[k]: k, c, ck, ch, cc, qu, que (5)

piquant (12)
fecund
inchoate
accolade
desecrate
ransack (11)
concomitant
succor
desiccate
baroque
hierarchy
anachronism
inculcate
peccadillo
cairn
skirmish (10)
recant
chimerical
archaic

[kw] (subcategory of k): qu (6)

equanimity
querulous
loquacious
acquiesce
ubiquitous
soliloquy
equivocal
magniloquent

[g]: g, gg, gu, gh, gue (7)	
polyglot	mitigate
epilogue	aghast (8)
litigation	choreography
configuration	elicit
aggregate (12)	guest (4)
egress	
vagary	
enigma	
gratuitous	
abrogate	

[ks]: x, xc, cc (8)
matrix
quixotic
accidentally (6)
excoriate
execrate
prolix
exception (6)
expiate

[gz]: x (9)
exemplary
exonerate
exacerbate

Sibilants

[s]: s, se, c, ss, ce, sc, z, ps (10)

perverse (12)
dissident
ascetic
nemesis
psychosis (11)
sonorous
irascible
incipient
rescind
miscegenation
assiduous
neoscience
esoteric
vicissitude
apostasy
decimate
salacious
lassitude
waltz (6)
nuance (12)

[z]: s, z, se, zz, ze, ss (11)

cataclysm
schism
analyze (8)
commiserate
buzzard (7)
risible
amortize
guise (11)
paroxysm
dispossess (9)
cognizant

[ʃ]: sh, ti, s, ch, ci, c, ssi, ss, si, xi (12)

voracious
pernicious
chauvinism
dimension (8)
adulation
satiate
cynosure
auspicious
fetish
factious
permutation
noxious
propitiate
remission
issue (6)
vitiate
fractious
oceanographer (6)

[ʒ]: si, s, g, x (gʒ) (13)

montage
luxury (6)
collusion
usury
persiflage
ménage
entourage

[tʃ]: ch, tch, t (14)

penchant
fatuous
trenchant
dispatch (8)
truncheon
expostulate
importunate

[dʒ]: j, dge, ge, g, gg, dg, d, dj (15)

readjust (10)
jocund
panegyric
fidget (8)
egregious
dudgeon
vestige (12)
gesticulate
graduation (6)
intransigence
suggest (4)
misogyny
cajole
plangent
abjure
stringent
exigency
contingent

Fricatives

[f]: f, ff, gh, ph (16)

efficacy
neophyte
surfeit
aphorism
officious
filial
enough (2)
deficit
xenophobia
nefarious
cacophonous
artifice
philistine
refractory
mellifluous
ephemeral

[v]: ve, v, f (17)

variegated
invective
convivial
evanescent
servile
virulence
atavism
revile
travail
of (1)
parvenu

[θ]: th (18)

plethora
antithesis
misanthrope
epithet
antipathy
apotheosis
monolith

[ð]: th, the (19)

noteworthy (9)
scathing (12)
lithe (10)

[h]: h, wh (20)

homily
cohesive
whole (3)
hiatus
reprehensible
who (1)
hauteur
halcyon

Nasals

[m]: m, mb, mm, mn (21)

peremptory
untrammeled
succumb (12)
commensurate
ameliorate
promulgate
damnable (9)
imprecate
demur
masochistic
luminary
immutable
remonstrate
emolument
consummate
matriculate

[n]: n, nn, kn, gn, pn (22)

superannuated
clandestine
acknowledge (7)
connote
pneumatic (12)
benign
necromancy
anomaly
millennium
vernal
omniscient
enmity
sententious
innocuous
incarnate

[ŋ]: ng, n (23)

punctilious
enervating
adjunct
harbinger
unction
propinquity
malinger
succinct
extenuating

Glides

[w]: w, wh, u (24)

unwonted
overwhelm (9)
assuage
sanguine
caterwaul

[r]: r, rr, wr, rh (25)

wrangle (11)
forensic
rhetoric (12)
garrulous
predilection
veracity
corroborate
deprecate
soporific
irrevocable
histrionic
recondite

[ɔr] (subcategory of r): or, oor, our, ore, oar, ar, aur (26)

portend
award (5)
stentorian
floor (2)
decorum
concourse (12)
dilatory
dinosaur (4)
florid
hoarse (4)
moribund
torpid
purport
uxorious
encore (9)
abortive

[l]: l, ll, le, il, el, al, ell (27)

libel (11)
malevolent
privilege (6)
bellicose
palliative
fallible
inclement
lateral (10)
alacrity
lien
implacable
culpable
philology
pusillanimous
satellite (5)
calligraphy

[j]: y, i (28)

imbroglio
yield (6)

Grade 13

Vowels

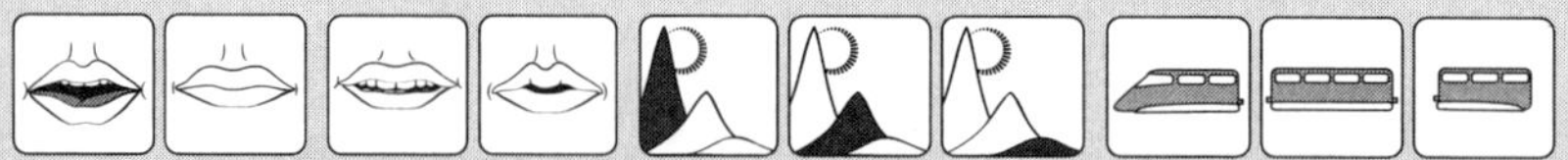

[i]: ee, e, y, e_e, ie, ea, ey, i, ei (29)

colleague (10)	effete
motley (12)	accede
etiology	imperious
absentee (9)	miscreant
deleterious	contumely
sartorial	ambient
amnesty	vitriolic
temerity	specious
credence	coterie
perceive (9)	aborigine

[ɪ]: i, y, ui (30)

declivity	implicit
inimical	syllogism
sycophant	perfidious
biscuit (6)	
vertiginous	
contiguous	
apocalyptic	
indigenous	
itinerant	
debility	

[ɛ]: ai, e, ea, ie, a, ae, hei (31)

polemic	irreparable
clemency	sedulous
aerial (7)	penury
mendacity	defection
unfriendly (4)	questionnaire (8)
zealot	equitable
dissemble	encomium
ancillary	vested
heiress (7)	carrion

[æ]: a, ai (32)

flaccid	facile
pandemic	
sporadic	
machination	
didactic	
masticate	
captious	
plaid (5)	
raconteur	

[ɑ] & [ɔ]: a, o, al, ough, aw, augh, u, au, oa (33)

trauma	incarcerate
tawdry (10)	laconic
broad (5)	oligarchy
distraught (11)	progeny
vaunt	dichotomy
altercation	autonomous
glaucous	qualm (10)
fought (4)	
culminate (12)	
improvident	

[ʊ]: oo, oul, u, ou (34)

ebullience
contour (9)
childhood (6)
should've (6)

[u]: ue, o, ew, oo, oe, ui, ou, u_e, u, eu (35)	
recruit (8)	elucidate
accouterments	milieu
fruition	construe
deciduous	lucid
certitude	renew (7)
suitable (6)	approval (6)
canoe (3)	
abstruse	
moot (12)	
ablution	
insouciance	

[ɝ] & [ɚ]: er, or, ir, ear, ur, urr, irr, ure, ar, our, err (36)	
onerous	interim
conferred (6)	squirrel (1)
interpolate	titular
covert	cursory
nourish (8)	precursor
timorous	dearth
circumspect	scurrilous
extirpate	avarice
appurtenances	saturnine
discomfiture	inexorable
insurgent	insuperable

[ʌ] & [ə] : u, o, o_e, a, e, ou, oo, i (37)		
truculent	hegemony	analogous
commodious	pundit	unctuous
blood (3)	comely (12)	affidavit (12)
flux	sublimate	
entrepreneur	covenant	
abstemious	ambivalence	
chicanery	ineluctable	
couples (6)	penumbra	
conjugal	calumny	
arrant	recumbent	

Diphonemes

[eɪ] & [e]: ay, a, a_e, ai, ey, eigh, aigh, ea, ei (38)

fray (9)
rapacious
abstain (11)
abnegate
straighten (5)
placate
surveyed (7)
vacate (9)
freight (4)
capacious
laity
break (2)
heinous
satrap
abeyance (12)
expatiate
denigrate

[oʊ] & [o]: o, o_e, oa, ow, oe, ou, ough, eau (39)

virago
hallowed (12)
cogent
encroach
quotidian
disown (7)
comatose
opiate
erode (12)
cantaloupe (7)
chateau (7)
parsimonious
thoroughbred (7)
procreate
hoe (5)
copious

[aʊ]: ou, ow, ough (40)

recount (12)
renown (9)
overpower (9)
drought (5)
surmount (11)

[aɪ]: y, i, i_e, igh, ie, ai, ye, y_e, ui (41)

lye (7)
rife
misguided (9)
stipend
belie (10)
demise
finite
hyperbole
iconoclast
geyser (7)
benighted
mollify
rhyme (5)
aisle (5)
diatribe
supine

[oɪ]: oy, oi (42)

foyer (8)
poignant (12)
convoy (8)
invoice (11)

[ju]: you, u_e, ew, ue, eau, u, uu, eu (43)

you'll (3)
eulogy
impunity
bucolic
vacuum (5)
effusive
pewter (9)
transmute
euphonious
imbue (12)
you (K)
erudite
pecuniary
corpulent
beauty (3)
emulate
munificent
cupidity

Index

Note: Page numbers in **bold** reference non-text material.

D

E

F

G

H

I

J

K

N

O

P

T

U

V